J. WESTON
WALCH
PUBLISHER

top SHELF

CALCULUS

MW00710925

$$\frac{dv}{dt} = 4\pi r^2 \frac{dr}{dt}$$

Joseph Caruso

User's Guide
to
Walch Reproducible Books

Purchasers of this book are granted the right to reproduce all pages.

This permission is limited to a single teacher, for classroom use only.

Any questions regarding this policy or requests to purchase further reproduction rights should be addressed to

Permissions Editor
J. Weston Walch, Publisher
321 Valley Street • P.O. Box 658
Portland, Maine 04104-0658

1   2   3   4   5   6   7   8   9   10

ISBN 0-8251-4619-4

Copyright © 2003
J. Weston Walch, Publisher
P.O. Box 658 • Portland, Maine 04104-0658
walch.com

Printed in the United States of America

Contents

To the Teacher

Creative problem solving, precise reasoning, effective communication, and alertness to the reasonableness of results are some of the essential areas of mathematics that educators have specified as necessary in the development of all students to function effectively in this century. It is in the spirit of the aforementioned competencies that *Top Shelf Math* is offered to teachers of mathematics.

Top Shelf Math is intended to help students become better problem solvers. The problems presented in this format are challenging, interesting, and can easily be blended into the teaching styles and strategies of teachers who are seeking supplementary problems that support and enhance the curriculum being taught.

In general, the topics selected for each of the major content areas of *Top Shelf Math* are typical of those found in the curricula of similarly named courses offered at the high school and early college level. All of the problems presented can be used by the teacher to help his or her students improve their problem-solving skills without slowing the pace of the course in which the students are enrolled.

Other areas of utilization of the problems presented in *Top Shelf Math* could be by teachers to help prepare their students for achievement tests and advanced placement tests. Math team coaches will find these problems especially useful as they prepare their students for math competitions. It is recommended that students save the problems and solutions presented in *Top Shelf Math* because they provide a rich resource of mathematical skills and strategies that will be useful for preparing to take standardized tests and enrollment in future mathematics courses.

Mathematical thought along with the notion of problem solving is playing an increasingly important role in nearly all phases of human endeavor. The problems presented in *Top Shelf Math* help provide the teacher with a mechanism for the students to witness a variety of applications in a wide sphere of real-life settings.

In problems requiring a calculator solution, it is recommended that only College Board-approved calculators be used. In addition, some problems will suggest that a calculator not be used and that the solution will require an algebraic procedure.

The approach for solving the problems presented in *Top Shelf Math* is consistent with emphasis by national mathematics organizations for reform in mathematics teaching and learning, content, and application by taking advantage of today's technological tools that are available to most if not all high school and college students enrolled in similarly named courses. *Top Shelf Math* provides the teacher with a balance of using these tools as well as well-established approaches to problem solving.

We hope that you will find the problems useful as general information as well as in preparation for higher-level coursework and testing. For additional books in the *Top Shelf Math* series, visit our web site at walch.com.

 **INSTRUCTION**

Limits and Continuity

> ## It is important to have a good working knowledge of limits.

The notion of a limit is fundamental to the study of calculus. It is important to have a good working knowledge of limits as you progress to studying other topics in calculus. In addition to having a good working knowledge of the notion of a limit, you must also draw upon a solid background of algebraic concepts and skills to be proficient in doing limit problems.

The number L is the **limit of the function** $f(x)$ as x approaches c if, as the values of x get arbitrarily close (but not equal) to c, the values of $f(x)$ approach (or equal) L.

$$\lim_{x \to c} f(x) = L$$

In order for $\lim\limits_{x \to c} f(x)$ to exist, the values of f must tend to the same number L as x approaches c from either the left or the right. It can be written $\lim\limits_{x \to c^-} f(x) = L$ for the **left-hand limit** of f at c (as x approaches c through values *less* than c) and $\lim\limits_{x \to c^+} f(x) = L$ for the **right-hand limit** of f at c (as x approaches c through *greater* than c).

The function $f(x)$ is said to *become infinite* (positively or negatively) as x approaches c if $f(x)$ can be made arbitrarily large (positively or negatively) by taking x sufficiently close to c. It can be written $\lim\limits_{x \to c} f(x) = +\infty$ (*or* $\lim\limits_{x \to c} f(x) = -\infty$). Because for the limit to exist, it must be a finite number, this definition can be extended to include x approaching c from the left or from the right.

The line $y = b$ is a horizontal asymptote of the graph of $y = f(x)$ if

$$\lim_{x \to \infty} f(x) = b \text{ or } \lim_{x \to -\infty} f(x) = b.$$

1

> **If a function is continuous over an interval, it can be drawn without lifting pencil from paper.**

If d, k, R, T, V, and W are finite numbers and if

$$\lim_{x \to d} f(x) = R, \quad \lim_{x \to d} g(x) = T, \quad \lim_{x \to \infty} f(x) = V,$$

$$\lim_{x \to \infty} g(x) = W, \text{ then}$$

$$\lim_{x \to d} kf(x) = kR \qquad\qquad \lim_{x \to \infty} kf(x) = kV$$

$$\lim_{x \to d} [f(x) + g(x)] = R + T \qquad\qquad \lim_{x \to \infty} [f(x) + g(x)] = V + W$$

$$\lim_{x \to d} [f(x)g(x)] = RT \qquad\qquad \lim_{x \to \infty} [f(x)g(x)] = VW$$

$$\lim_{x \to d} \frac{f(x)}{g(x)} = \frac{R}{T} \text{ (if } g(x) \neq 0) \qquad\qquad \lim_{x \to \infty} \frac{f(x)}{g(x)} = \frac{V}{W} \text{ (if } g(x) \neq 0)$$

$$\lim_{x \to d} k = k \qquad\qquad \lim_{x \to \infty} k = k$$

To find the $\lim\limits_{x \to \infty} \dfrac{P(x)}{Q(x)}$, where $P(x)$ and $Q(x)$ are polynomials in x, we can divide both numerator and denominator by the highest power of x that occurs and use the fact that $\lim\limits_{x \to \infty} \dfrac{1}{x} = 0$.

If a function is **continuous** over an interval, it can be drawn without lifting pencil from paper. The graph has no holes, breaks, or jumps on the interval. All *three* of the following statements must be true for $f(x)$ to be continuous at $x = c$.

(i) $f(c)$ must exist. (ii) $\lim\limits_{x \to c} f(x)$ must exist. (iii) $\lim\limits_{x \to c} f(x) = f(c)$.

A function that is not continuous at $x = c$ is said to be discontinuous at that point, which is called the *point of* **discontinunity.**

π TRY IT | **Practice Activities**

1. (a) Find and prove $\displaystyle\lim_{x \to 2} \frac{3}{x+1}$.

    (b) Using the results of the proof, suppose $\varepsilon = 0.005$, find δ (correct to thousandths) and the interval that insures that $-\varepsilon < f(x) - L < \varepsilon$.

2. Find the limits as specified. Show all work.

    (a) $\displaystyle\lim_{x \to 0} \frac{\dfrac{1}{2+x} - \dfrac{1}{2}}{x}$

    (b) $\displaystyle\lim_{x \to 2} \frac{x^3 - 8}{x^2 - 4}$

    (c) $\displaystyle\lim_{x \to 0} \frac{(\csc 5x)(\sec 8x)}{(\csc 8x)(\sec 3x)}$

    (d) $\displaystyle\lim_{x \to 0} \frac{x + \sin 2x}{x}$

    (e) $\displaystyle\lim_{x \to 0} \frac{\cos 2x}{\sin x - \cos x}$

3. Discuss the continuity of $f(x) = \begin{cases} -x-1, & -2 < x < 0 \\ 2, & x = 0 \\ -x, & 0 < x < 2 \\ 0, & x = 2 \\ x-4, & 2 < x \le 4 \end{cases}$

4. What value should be assigned to b to make the function

$$f(x) = \begin{cases} x^2 - 1, & x < 3 \\ 2bx, & x \ge 3 \end{cases}$$

continuous at $x = 3$? Sketch a complete graph of $f(x)$ for this value of b.

5. The function defined by $f(x) = \begin{cases} x^3, & x \le 2 \\ 2x + A, & x > 2 \end{cases}$ is continuous at $x = 2$. Find A.

6. Find the limits as specified. Show all work.

(a) $\lim\limits_{x \to \infty} \dfrac{3 + x}{4 + x + x^2}$

(b) $\lim\limits_{x \to \infty} \dfrac{4x^4 + 5x + 1}{37x^3 - 9}$

(c) $\lim\limits_{x \to \infty} \dfrac{x^3 - 4x^2 + 7}{3 - 6x - 2x^3}$

4

 **INSTRUCTION**

The Derivative

The **derivative** of a function expresses its rate of change with respect to an **independent variable.** The derivative is also the slope of the **tangent line to the curve.**

> The derivative of a function expresses its rate of change with respect to an independent variable.

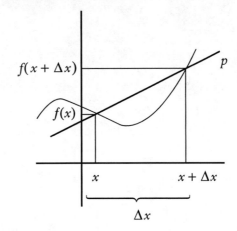

The notation used for the derivative of the function f is f' and we have the following definition:

$$f'(x) = \lim_{\Delta x \to 0} \frac{f(x + \Delta x) - f(x)}{\Delta x}$$

When doing the algebraic manipulation to find the derivative, it may be more convenient to use h for Δx.

π   **TRY IT**   **Practice Activities**

1. Using the definition of the derivative $f'(x) = \lim\limits_{h \to 0} \dfrac{f(x+h) - f(x)}{h}$, find the derivatives of the following functions:

(a) $f(x) = \dfrac{2x - 1}{x + 2}$

(b) $f(x) = \sqrt{3x - 1}$

(c) $f(x) = 3x^4$

 **INSTRUCTION**

Numerical Derivatives

Some graphing calculators can compute derivatives by using rules of differentiation. Some can approximate derivatives by applying a numerical method. This section assumes the use of a graphing calculator that can do the latter and accurately approximate the derivative of any differentiable function at most values in its domain. The numerical derivative of a function at a point will be denoted by the expression: NDER $f(a)$ or NDER $(f(x), a)$ and the numerical derivative (as a function) will be denoted by the expression: NDER $f(x)$.

Many graphing calculators use the symmetric difference quotient $\dfrac{f(a+h) - f(a-h)}{2h}$ to determine NDER $(f(x), a)$. Many graphing calculators allow the user to specify how close to 0 they want h to be for the approximation of $f'(a)$.

π **TRY IT** **Practice Activities**

1. Using the symmetric difference quotient for $f'(x_0) = \dfrac{f(x_0 + h) - f(x_0 - h)}{2h}$, to how many places is the symmetric difference accurate when it is used to approximate $f'(0)$ for $f(x) = 4^x$ and $h = 0.08$?

2. Using the symmetric difference quotient of Problem 1 and $f(x) = x^4$, find $f'(1)$ and $h = 0.01$.

    Compare your results with the graphing calculator's results by using NDER $(f(x), a)$. Using NDER $(f(x), a)$, find the equation of the tangent line to the curve at $x = 1$. Sketch $f(x)$ and the tangent line on the same axes.

8

 **INSTRUCTION**

Rules for Differentiation

The limit definition for a derivative can be used to find derivatives of functions. However, this is often a tedious process and "differentiation rules" are available to allow you to find derivatives without direct use of the limit definition. Rules for finding the derivative of a function can be found below. It is essential that you gain familarity with using them. In these formulas, *a* and *n* are constants and *u* and *v* are **differentiable functions** of *x*.

$$\frac{d}{dx}a = 0 \qquad\qquad \frac{d}{dx}(u + v) = \frac{d}{dx}u + \frac{d}{dx}v$$

$$\frac{d}{dx}au = a\frac{du}{dx} \qquad\qquad \frac{d}{dx}(uv) = u\frac{dv}{dx} + v\frac{du}{dx}$$

$$\frac{d}{dx}u^n = nu^{n-1}\frac{du}{dx} \qquad\qquad \frac{d}{dx}\left(\frac{u}{v}\right) = \frac{v\frac{du}{dx} - u\frac{dv}{dx}}{v^2}$$

  Practice Activities

1. Find y'.

    (a) $y = \sqrt{x^2 + 2x - 1}$

    (b) $y = \dfrac{x}{\sqrt{1 - x^2}}$

    (c) $y = \tan^{-1}\dfrac{x}{2}$

    (d) $y = \ln(\sec x + \tan x)$

    (e) $y = \dfrac{e^x - e^{-x}}{e^x + e^{-x}}$

2. Find $\dfrac{dy}{dx}$.

    (a) $y = x^2 \sin\dfrac{1}{x}$

    (b) $y = \dfrac{1}{2\sin 2x}$

    (c) $y = \sec^2\sqrt{x}$

    (d) $y = x \ln^3 x$

    (e) $y = \sin^{-1}x - \sqrt{1 - x^2}$

3. $f(x) = (2x - 1)^{100}$. Find a, b, c, d for $f^{10}(x) = \dfrac{a!}{b!}(2^c)(2x - 1)^d$.

4. If $y = x^2 + x$, find the derivative of y with respect to $\dfrac{1}{1 - x}$.

 **INSTRUCTION**

Velocity, Speed, and Other Rates of Change

> **Applications involving rates of change can be found in a wide variety of fields.**

Derivatives provide the mathematics we need to understand the way in which things change in the world around us. One important use is to determine the rate of change of one variable with respect to another. Applications involving rates of change can be found in a wide variety of fields. This section looks at a few of these applications.

The function *s* that gives the position of an object as a function of time is called the position function. The velocity of the object at time *t* is the derivative of the position function, and the **acceleration** of the object at time *t* is the derivative of the velocity. The position is expressed in distance units; the velocity is expressed in distance units per time units; the acceleration is expressed in distance units per unit of time squared units.

$$\text{If } s = s(t), \text{ then } v(t) = s'(t) = \frac{ds}{dt} \text{ and } v'(t) = \frac{dv}{dt}.$$

Marginal cost is the rate of change of the total production cost with respect to the number of units produced.

$$\text{Marginal cost} = MC = C'(x)$$

where $C(x)$ is the total production cost of producing *x* units.

Marginal revenue is the rate of change of the total revenue with respect to the number of units sold.

$$\text{Marginal revenue} = MR = R'(x)$$

where $R(x)$ is the total revenue derived from selling *x* units.

11

π   TRY IT   **Practice Activities**

1. The total cost in dollars, $C(x)$, for manufacturing x units is given by $C(x) = 2x^3 - 5x^2 + 8$, $x \geq 2$.

   Find

   (a) the marginal cost equation

   (b) the marginal cost at $x = 10$ level of production

   (c) interpret the answer to part (b)

2. The total cost of producing x units of some commodity is given by the cost function $C(x) = 20 + 2x + 0.01x^2$, $x \geq 0$.

   Find

   (a) the equation for marginal cost

   (b) the marginal cost at $x = 100$ and $x = 200$

   (c) interpret the answers to part (b)

   (d) the cost of producing 100 units and 101 units

   (e) the cost of producing the 101st unit

3. A ball is tossed upward from the top of a tower. The function that describes the height above the ground of the ball after t seconds is $H(x) = -16t^2 + 64t + 80$ ft. Find

   (a) the instantaneous velocity at $t = 1$

   (b) the height of the ball after 4 seconds

   (c) the velocity of the ball after 4 seconds

   (d) the initial velocity with which the ball was tossed

   (e) the maximum height reached by the ball

   (f) when the ball will hit the ground

   (g) the velocity of the ball at the instant it hits the ground

4. Last year in Worcester County, the number of manufactured housing sales was found to be 1,757. At that time, it was projected that t months from then, the number of manufactured housing sales for the county would be $N(t) = \frac{1}{3}t^3 - 6t^2 + t + 1{,}757$.

   (a) Determine the formula for the instantaneous rate of change for the number of manufactured housing sales with respect to time.

   (b) At what rate was the number of housing sales changing with respect to time six months from then? Were the sales increasing or decreasing at that point in time?

   (b) At what rate was the number of housing sales changing with respect to time twelve months from then? Were the sales increasing or decreasing at that point in time?

5. The number of gallons of water in a tank t minutes after the tank has started to drain is given by $Q(t) = 200(30 - t)^2$. How fast is the water running out at the end of ten minutes? What is the average rate at which the water flows out during the first ten minutes?

 **INSTRUCTION**

Interpreting Graphs of Functions and Their Derivatives

If the derivative of $y = f(x)$ exists at $P(x_1, y_1)$, then the **slope of the curve** at P (which is defined to be the slope of the tangent to the curve at P) is $f'(x)$, the derivative of $f(x)$ at $x = x_1$. Any c in the **domain** of f such that either $f'(c) = 0$ or $f'(c)$ is undefined is called a **critical point** or critical value of f. If f has a derivative everywhere, we find the critical points by solving the equation $f'(x) = 0$.

A function is said to be *increasing/decreasing* at $P(x_1, y_1)$ if its derivative $f'(x_1)$ is *positive/negative*. To find intervals over which $f(x)$ *increases/decreases*—that is, over which the curve *rises/falls*—compute $f'(x)$ and determine where it is *positive/negative*.

The curve $y = f(x)$ has a *local* (or *relative*) *maximum/minimum* at a point where $x = c$ if $f(c) \geq f(x) / f(c) \leq f(x)$ for all x in the immediate neighborhood of c. If a curve has a local *maximum/ minimum* at $x = c$, then the curve changes from *rising/falling* to *falling/rising* as x increases through c.

If $f(c)$ is either a local maximum or a local minimum, the $f(c)$ is called a *local extreme value* or *local extremum* (the plural of *extremum* is *extrema*).

The *global* or *absolute maximum/minimum* of a function on $[a, b]$ occurs at $x = c$ if $f(c) \geq f(x) / f(c) \leq f(x)$ for all x on $[a, b]$. $f(c)$ is called an extreme value of f on $[a, b]$ if it is either the global maximum or the global minimum of f on $[a, b]$.

A curve is said to be **concave upward/downward** at a point $P(x_1, y_1)$ if the curve lies *above/below* its tangent. If $y'' > 0 / y'' < 0$ at P, the curve is concave *up/down*. A **point of inflection** is a point where the curve changes its concavity from upward to downward or from downward to upward. The point of inflection is found by solving $f''(x) = 0$.

The following procedure is suggested when seeking to determine maximum, minimum, or inflection points of a curve. This procedure also suggests how to find intervals where the curve is increasing, decreasing, concave upward, or concave downward.

> **A point of inflection is a point where the curve changes its concavity from upward to downward or from downward to upward.**

14

1. Find y' and y''.

2. Find all critical points of y by solving $y' = 0$. At each of these x's, the tangent to the the curve is horizontal.

3. First-Derivative Test. If c is a critical value of f that is continuous on an open interval containing c, and if f is differentiable on the interval, except possibly at c, then

   $f(c)$ is   a *relative minimum* if f' changes from negative to positive at c

   a *relative maximum* if f' changes from positive to negative at c

   neither if f' doesn't change sign

4. Second-Derivative Test. If $f'(c) = 0$ and the second derivative of f exists on an open interval containing c, then

   $f(c)$ is a *relative minimum* if $f''(c) > 0$

   $f(c)$ is a *relative maximum* if $f''(c) < 0$

5. If $(c, f(c))$ is a point of inflection of the graph of f, then either $f''(c) = 0$ or f'' is undefined at $x = c$.

6. If $f'(x) > 0$ for all x on (a, b), then f is increasing on (a, b).

   If $f'(x) < 0$ for all x on (a, b), then f is decreasing on (a, b).

   If $f''(x) > 0$ for all x on (a, b), then f is concave upward on (a, b).

   If $f''(x) < 0$ for all x on (a, b), then f is concave downward on (a, b).

If a function f is differentiable on $[a, b]$, then f is also continuous on $[a, b]$ and f attains both a (global) maximum and a (global) minimum on $[a, b]$. To find these values, solve the equation $f'(x) = 0$ for the critical points on the interval $[a, b]$, then evaluate f at each of those and also at $x = a$ and $x = b$. The largest value of f obtained is the global maximum and the smallest the global minimum.

 **Practice Activities**

1. For $f(x) = \dfrac{x^4 + 1}{x^2}$, find any maximum, minimum, points of inflection, intervals where the function may be increasing or decreasing, concave up, or concave down. Sketch the curve.

2. For $f(x) = x^3 - 5x^2 + 3x + 6$, find any maximum, minimum, points of inflection, intervals where the function may be increasing or decreasing, concave up, or concave down. Sketch the curve.

3. For $f(x) = \dfrac{x^2 + 1}{x^2 - 4}$, find any maximum, minimum, points of inflection, intervals where the function may be increasing or decreasing, concave up, or concave down. Sketch the curve.

4. For $f(x) = \dfrac{2x + 1}{x - 1}$, find any maximum, minimum, points of inflection, intervals where the function may be increasing or decreasing, concave up, or concave down. Sketch the curve.

5. Determine the constants a, b, and c in order for the function $f(x) = x^3 + ax^2 + bx + c$ to have a relative maximum at $x = 1$ and a point of inflection at $x = 2$. Find a relative minimum of the function given the found values of a and b, provided that $f(0) = 6$.

16

 **INSTRUCTION**

Derivatives and Integrals of Trigonometric Functions

> **The derivative is also the slope of the tangent line to the curve.**

As previously stated in "The Derivative," the derivative of a function expresses its rate of change with respect to an independent variable. The derivative is also the slope of the tangent line to the curve. Basic techniques (such as product rule, quotient rule, power rule) for finding derivatives were provided in "Rules for Differentiation" and are repeated here for convenience. Again in these formulas, a and n are constants, and u and v are differentiable functions of x.

$$\frac{d}{dx}a = 0$$

$$\frac{d}{dx}au = a\frac{du}{dx}$$

$$\frac{d}{dx}u^n = nu^{n-1}\frac{du}{dx} \text{ (power rule)}$$

$$\frac{d}{dx}(u + v) = \frac{d}{dx}u + \frac{d}{dx}v$$

$$\frac{d}{dx}(uv) = u\frac{dv}{dx} + v\frac{du}{dx} \text{ (product rule)}$$

$$\frac{d}{dx}\left(\frac{u}{v}\right) = \frac{v\frac{du}{dx} - u\frac{dv}{dx}}{v^2} \text{ (quotient rule)}$$

Methods or formulas for taking derivatives of trigonometric functions will now be expanded to the techniques listed below:

$$\frac{d}{dx}\sin u = \cos u \frac{du}{dx} \qquad \frac{d}{dx}\cot u = -\csc^2 u \frac{du}{dx}$$

$$\frac{d}{dx}\cos u = -\sin u \frac{du}{dx} \qquad \frac{d}{dx}\sec u = \sec u \tan u \frac{du}{dx}$$

$$\frac{d}{dx}\tan u = \sec^2 u \frac{du}{dx} \qquad \frac{d}{dx}\csc u = -\csc u \cot u \frac{du}{dx}$$

The **antiderivative** or **indefinite integral** of a function $f(x)$ is a function $F(x)$ whose derivative is $f(x)$. Because the derivative of a constant equals zero, the antiderivative of $f(x)$ is not unique: that is, if $F(x)$ is an integral of $f(x)$, then so is $F(x) + C$, where C is any constant. The arbitrary constant C is called the **constant of integration**. The indefinite integral of $f(x)$ is written

$$\int f(x)\,dx = F(x) + C \ \ if \ \frac{d(F(x))}{dx} = f(x).$$ The function $f(x)$ is

called the **integrand**. Familarity with the following fundamental integration formulas is essential. Again in these formulas, a, k, and n are constants and $f(x)$, $g(x)$, u, and v are differentiable functions of x.

$$\int kf(x)\,dx = k\int f(x)\,dx \qquad \int [f(x) + g(x)]\,dx = \int f(x)\,dx + \int g(x)\,dx, \qquad \int u^n\,du = \frac{u^{n+1}}{n+1} + C$$

The integration formulas listed above are used in conjunction with the integration formulas for trigonometric functions listed below:

$$\int \cos u \, du = \sin u + C \qquad\qquad \int \csc^2 u \, du = -\cot u + C$$

$$\int \sin u \, du = -\cos u + C \qquad\qquad \int \sec u \tan u \, du = \sec u + C$$

$$\int \tan u \, du = \ln|\sec u| + C \ or -\ln|\cos u| + C \qquad \int \csc u \cot u \, du = -\csc u + C$$

$$\int \cot u \, du = -\ln|\csc u| + C \ or \ \ln|\sin u| + C \qquad \int \sec u \, du = \ln|\sec u + \tan u| + C$$

$$\int \sec^2 u \, du = \tan u + C \qquad\qquad \int \csc u \, du = \ln|\csc u - \cot u| + C$$

π TRY IT   Practice Activities

1. Given $\dfrac{dy}{dx}$, find y and express it in simplest form.

   (a) $\dfrac{dy}{dx} = 3\cos^2 x - 3\sin^2 x$

   (b) $\dfrac{dy}{dx} = x\cot(3x^2)$

2. Let $f'(x) = \sin(\pi x)$ and $f(0) = 0$. Find $f(-1)$.

3. If $y = \csc(t + \sqrt{t})$, find $y'(1)$ in radians correct to thousandths.

4. (a) $\displaystyle\int (\tan\theta - 1)^2\, d\theta =$

   (b) $\displaystyle\int \sec^{\frac{2}{3}} x\ \tan x\, dx$

5. Find $\dfrac{dy}{dx}$ of the following examples:

   (a) $y = x^2 \sin\dfrac{1}{x}$

   (b) $y = \dfrac{1}{2 \sin 2x}$

   (c) $y = \sec^2 \sqrt{x}$

   (d) $x = \cos^3 \theta$ and $y = \sin^3 \theta$

   (e) $\sin x - \cos y - 2 = 0$

   (f) $\sin(xy) = x$

6. If $y = a \sin ct + b \cos ct$, where a, b, and c are constants, then find $\dfrac{d^2 y}{dx^2}$.

7. $\displaystyle\int \dfrac{36 \tan^2 x \sec^2 x}{\left(6 + \tan^3 x\right)^2}\,dx$

8. (a) $\displaystyle\int \sec^4 x\,dx$

   (b) $\displaystyle\int \dfrac{\cos \sqrt{x}}{\sqrt{x}}\,dx$

   (c) $\displaystyle\int \cos^2 x\, \sin^3 x\,dx$

 **INSTRUCTION**

Derivatives and Integrals of Exponential Functions

In calculus, the natural (or convenient) choice for a base is the irrational number e.

Exponential functions are of the form $f(x) = a^x$, where a is a positive number (called the base) and $a \neq 1$. In calculus, the natural (or convenient) choice for a base is the irrational number e, for which the decimal approximation is $e \approx 2.71828182846....$ The natural number e is defined by the limit: $e = \lim_{x \to 0} (1 + x)^{\frac{1}{x}}$. Basic techniques (such as product rule, quotient rule, power rule) for finding derivatives that were provided in previous sections are repeated here for convenience.

Again in these formulas, a and n are constants, and u and v are differentiable functions of x. Included here is the natural exponential function $f(x) = e^x$.

$$\frac{d}{dx}a = 0$$

$$\frac{d}{dx}au = a\frac{du}{dx}$$

$$\frac{d}{dx}u^n = nu^{n-1}\frac{du}{dx} \text{ (power rule)}$$

$$\frac{d}{dx}(u + v) = \frac{d}{dx}u + \frac{d}{dx}v$$

$$\frac{d}{dx}(uv) = u\frac{dv}{dx} + v\frac{du}{dx} \text{ (product rule)}$$

$$\frac{d}{dx}\left(\frac{u}{v}\right) = \frac{v\frac{du}{dx} - u\frac{dv}{dx}}{v^2} \text{ (quotient rule)}$$

Methods or formulas for taking derivatives of exponential functions will now be expanded to the techniques listed below:

$$\frac{d}{dx}e^u = e^u\frac{du}{dx} \qquad \frac{d}{dx}a^u = (\ln a)a^u\frac{du}{dx}$$

As was the case for derivatives, the basic integration formulas are repeated here for convenience. Again in these formulas, a and n are constants, and u and v are differentiable functions of x. Included here is the natural exponential function $f(x) = e^x$.

$$\int kf(x)dx = k\int f(x)dx, \quad \int [f(x) + g(x)]dx = \int f(x)dx + \int g(x)dx, \quad \int u^n du = \frac{u^{n+1}}{n+1} + C$$

The integration formulas listed above are used in conjunction with the integration formulas for exponential functions listed below:

$$\int e^u du = e^u + C, \quad \int a^u du = \frac{a^u}{\ln a} + C \ (a > 0, a \neq 1)$$

π **TRY IT**   **Practice Activities**

1. (a) $\int \dfrac{x + e^x}{xe^x}\,dx$

    (b) $\int \dfrac{\log(x^3 \times 10^x)}{x}\,dx$

    (c) $\int \dfrac{e^{2x}}{e^x - 3}\,dx$

2. If $f(x) = 2^{8x^3 + 1}$, then find the exact value of $f'(0.5)$.

3. Let $f(x) = e^{2bx}$ and $g(x) = e^{2ax}$. Find the value of b such that $\dfrac{dy}{dx}\left(\dfrac{f(x)}{g(x)}\right) = \dfrac{f'(x)}{g'(x)}$.

4. If $f(x) = 3e^{kx}$ and $\dfrac{f'(x)}{f(x)} = -\dfrac{a}{b}$, then $k =$

5. (a) $\int e^{2\theta} \sin e^{2\theta}\,d\theta$

    (b) $\int \dfrac{e^{2x}}{1 + e^x}\,dx$

    (c) $\int e^{2\ln x}\,dx$

 **INSTRUCTION**

Derivatives and Integrals of Logarithmic Functions

Because the natural exponential function $f(x) = e^x$ is continuous and increasing on the entire real number line, it must possess an inverse function. This inverse function is called the natural logarithmic function. The domain of the natural logarithmic function is the set of positive real numbers and is defined as follows:

$$\ln x = b \text{ if and only if } e^b = x$$

Just as the natural logarithmic function was defined as the inverse of the natural exponential function, the **logarithmic function** to any positive base $a \neq 1$ is the inverse of the exponential function $f(x) = a^x$. Remember, when $a = 10$, the function given by $\log_{10} x$ is called the **common logarithmic function**.

Basic techniques (such as product rule, quotient rule, power rule) for finding derivatives that were provided in previous sections are repeated here for convenience. Again in these formulas, a and n are constants, and u and v are differentiable functions of x. Included here is the natural logarithmic function $f(x) = \ln x$ and the common logarithmic function $f(x) = \log_{10} x$, or more simply stated, $f(x) = \log x$.

> **Remember, when $a = 10$, the function given by $\log_{10}x$ is called the common logarithmic function.**

$$\frac{d}{dx}a = 0$$

$$\frac{d}{dx}au = a\frac{du}{dx}$$

$$\frac{d}{dx}u^n = nu^{n-1}\frac{du}{dx} \text{ (power rule)}$$

$$\frac{d}{dx}(u+v) = \frac{d}{dx}u + \frac{d}{dx}v$$

$$\frac{d}{dx}(uv) = u\frac{dv}{dx} + v\frac{du}{dx} \text{ (product rule)}$$

$$\frac{d}{dx}\left(\frac{u}{v}\right) = \frac{v\frac{du}{dx} - u\frac{dv}{dx}}{v^2} \text{ (quotient rule)}$$

The derivative rule for inverses: $g'(f(x)) = \dfrac{1}{f'(x)}$.

Methods or formulas for taking derivatives of logarithmic functions will now be expanded to the techniques listed below:

$$\frac{d}{dx}\ln u = \frac{1}{u} \cdot \frac{du}{dx}, \ u > 0, \qquad \frac{d}{dx}(\log_a u) = \frac{1}{(\ln a)u} \cdot \frac{du}{dx}$$

Method of Logarithmic Differentiation

1. For functions in the form of $y = u$, take the natural logarithm of both sides.

2. Use logarithmic properties to rid $\ln u$ of as many products, quotients, and exponents as possible.

3. Differentiate implicitly.

4. Solve for $\dfrac{dy}{dx}$.

5. Substitute for y.

As was the case for derivatives, the basic integration formulas are repeated here for convenience and extended to include integrals of the form $\log_a x$ and $\displaystyle\int \frac{du}{u}$. Again in these formulas, a and n are constants, and u and v are differentiable functions of x.

$$\int kf(x)\,dx = k\int f(x)\,dx, \quad \int [f(x) + g(x)]\,dx = \int f(x)\,dx + \int g(x)\,dx, \quad \int u^n\,du = \frac{u^{n+1}}{n+1} + C$$

The integration formulas listed above are used in conjunction with the integration formulas for exponential functions listed below:

To evaluate integrals involving base a logarithms, we convert them to natural logarithms and use an appropriate integration technique.

$$\int \frac{du}{dx} = \ln|u| + C$$

π **TRY IT**   **Practice Activities**

1. Find the exact value of $\displaystyle\int_{1}^{\sqrt{5}} \frac{\ln(x^2)}{x}\,dx$.

2. Find $\dfrac{dy}{dx}$ for each of the following equations:

   (a) $y = \ln[(x+1)(x+2)]$

   (b) $y = -\ln\left|\dfrac{1+\sqrt{1-x^2}}{x}\right|$

3. The solution of the equation $(x+1)\dfrac{dy}{dx} = x(y^2+1)$ is

4. A particle is moving on the curve $y = 2x - \ln 3x$ so that $\dfrac{dx}{dt} = -2$ at all times t. At the point $(1, 2)$, find $\dfrac{dy}{dt}$.

5. Assuming y is positive, find $\dfrac{dx}{dt}$ by the method of logarithmic differentiation if

   $y = \dfrac{x+5}{x\cos x}$.

6. (a) $\int \dfrac{(x-2)^3}{x^2}\,dx$

    (b) $\int \dfrac{x^3 - x - 1}{(x+1)^2}\,dx$

    (c) $\int \dfrac{(1 - \ln t)^2}{t}\,dt$

    (d) $\int \dfrac{2x - 1}{\sqrt{4x - 4x^2}}\,dx$

7. Find $\dfrac{dy}{dx}$ of the following:

    (a) $y = \ln(x\sqrt{x^2 + 1})$

    (b) $y = x\ln^3 x$

    (c) $x = \dfrac{1}{1 - t}$ and $y = 1 - \ln(1 - t)$, $t < 1$

8. If $f(x) = \ln x^3$, then $f''(3) =$

Derivatives and Integrals of Inverse Trigonometric Functions

The inverse trigonometric functions arise in problems that require finding angles from side measurements in triangles. They also provide antiderivatives for a wide variety of functions. As in previous sections, basic techniques for finding derivatives as well as integrals are provided.

> **The inverse trigonometric functions arise in problems that require finding angles from side measurements in triangles.**

$$\frac{d}{dx}a = 0$$

$$\frac{d}{dx}au = a\frac{du}{dx}$$

$$\frac{d}{dx}u^n = nu^{n-1}\frac{du}{dx} \text{ (power rule)}$$

$$\frac{d}{dx}(u + v) = \frac{d}{dx}u + \frac{d}{dx}v$$

$$\frac{d}{dx}(uv) = u\frac{dv}{dx} + v\frac{du}{dx} \text{ (product rule)}$$

$$\frac{d}{dx}\left(\frac{u}{v}\right) = \frac{v\frac{du}{dx} - u\frac{dv}{dx}}{v^2} \text{ (quotient rule)}$$

$$\frac{d}{dx}(\sin^{-1}u) = \frac{1}{\sqrt{1-u^2}}\frac{du}{dx}, -1 < u < 1$$

$$\frac{d}{dx}(\cos^{-1}u) = -\frac{1}{\sqrt{1-u^2}}\frac{du}{dx}, -1 < u < 1$$

$$\frac{d}{dx}(\tan^{-1}u) = \frac{1}{1+u^2}\frac{du}{dx}$$

$$\frac{d}{dx}(\cot^{-1}u) = -\frac{1}{1+u^2}\frac{du}{dx}$$

$$\frac{d}{dx}(\sec^{-1}u) = \frac{1}{|u|\sqrt{u^2-1}}\frac{du}{dx}, \ |u|>1$$

$$\frac{d}{dx}(\csc^{-1}u) = -\frac{1}{|u|\sqrt{u^2-1}}\frac{du}{dx}, \ |u|>1$$

$$\int kf(x)\,dx = k\int f(x)\,dx, \quad \int [f(x)+g(x)]\,dx = \int f(x)\,dx + \int g(x)\,dx, \quad \int u^n\,du = \frac{u^{n+1}}{n+1}+C$$

$$\int \frac{du}{\sqrt{1-u^2}} = \sin^{-1}u + C \qquad\qquad \int \frac{du}{1+u^2} = \tan^{-1}u$$

$$\int \frac{du}{u\sqrt{u^2-1}} = \int \frac{d(-u)}{(-u)\sqrt{u^2-1}} = \sec^{-1}|u| + C = \cos^{-1}\left|\frac{1}{u}\right| + C$$

π **TRY IT**   # Practice Activities

1. If $\tan^{-1}(2x) = \ln(y^2)$, find $\dfrac{dy}{dx}$ in terms of x and y.

2. Find $\dfrac{dy}{dx}$ for each of the following:

   (a) If $y = \tan(\sec^{-1}x)$, then $\dfrac{dy}{dx} =$

   (b) If $y = \tan(\cos^{-1}x)$, then $\dfrac{dy}{dx} =$

   (c) $y = \tan^{-1}\dfrac{x}{2}$

   (d) $y = \sin^{-1}x - \sqrt{1-x^2}$

3. (a) $\displaystyle\int \dfrac{dy}{\sqrt{6y-y^2}}$

   (b) $\displaystyle\int \dfrac{e^x}{3+e^{2x}}dx$

   (c) $\displaystyle\int \dfrac{\sqrt{x-1}}{x}dx$

4. (a) $\displaystyle\int \dfrac{x^2}{\sqrt{4-x^2}}dx$

   (b) $\displaystyle\int \dfrac{dx}{x\sqrt{4x^2+9}}$

INSTRUCTION

The Chain Rule

> **The chain rule says "Find the derivative of the 'outside' function first, then multiply by the derivative of the 'inside' one."**

Suppose we think of y as the composition function $f(g(x))$, where $y = f(u)$ and $u = g(x)$ are differentiable functions. Then

$$(f(g(x)))' = f'(g(x)) \cdot g'(x)$$

$$= f'(u) \cdot g'(x)$$

$$= \frac{dy}{du} \cdot \frac{du}{dx}$$

This is the *chain rule* that is used for differentiating the composition function. Basically it says "Find the derivative of the 'outside' function first, then multiply by the derivative of the 'inside' one."

π TRY IT   **Practice Activities**

1. Find $\dfrac{dy}{dx}$ of the following:

    (a) $y = \cos(\sqrt{3}x)$

    (b) $y = -\csc(x^2 + 7x)$

    (c) $y = \cos(\sin x)$

    (d) $y = \left(\dfrac{x}{5} + \dfrac{1}{5x}\right)^5$

2. Find $\dfrac{dy}{dx}$ of the following:

    (a) $y = x^3(2x - 5)^4$

    (b) $y = \left(\dfrac{1 + \cos x}{\sin x}\right)^{-1}$

    (c) $y = 2\sqrt{\csc x + \cot x}$

    (d) $y = (1 + \cos^2 7x)^3$

3. Show that the curves $y = \sin 2x$ and $y = -\sin\dfrac{x}{2}$ are orthogonal.

4. The temperature that approximates the average temperature (°F) on day t in Portland, Oregon, during a typical 365-day year is $y = 32\sin\left[\dfrac{2\pi}{365}(t - 88)\right] + 23$. On what day is the temperature increasing the fastest? About how many degrees per day is the temperature increasing when it is increasing at its fastest?

32

 **INSTRUCTION**

Implicit Differentiation

Implicit differentiation is the technique used to find a derivative when *y* is not defined explicitly in terms of *x* but is differentiable.

When a functional relationship between x and y is defined by an equation of the form $F(x, y) = 0$, we say that the equation defines *y implicitly* as a function of x. Some examples are $x^2 + y^2 = 16$ *and* $\cos(xy) = y^2 - 5$. Sometimes two (or more) explicit functions are defined by $F(x, y) = 0$. For example, $x^2 + y^2 = 16$ defines the two functions $y_1 = +\sqrt{16 - s^2}$ and $y_2 = -\sqrt{16 - x^2}$, the upper and lower halves, respectively, of the circle with the center at the origin and having a radius of 4. Each function is differentiable except at the points where $x = 3$ and $x = -3$. Implicit differentiation is the technique used to find a derivative when y is not defined explicitly in terms of x but is differentiable. The process involves differentiating both sides of the equation with respect to x, using appropriate formulas, and then solving for $\frac{dy}{dx}$.

33

π   TRY IT   **Practice Activities**

1. Find the slope of the line in the first quadrant that is tangent to the curve
    $y^3 + x^2 y^2 - 3x^2 = 9$ when $y = 2$.

2. If $2x^3 + 3xy + e^y = 6$, find the exact value of $\dfrac{dy}{dx}$ when $x = 0$.

3. Suppose $x^2 - xy + y^2 = 3$. Find $\dfrac{dy}{dx}$ at the point $(2a, -b)$.

4. Find the equations of the tangent and normal at $(1, -1)$ given the curve
    $4x^2 + 2xy - xy^3 = 0$.

 **INSTRUCTION**

Applications of the Derivative

Some of the applications that will be treated in this section include using the concept of the derivative in problems involving tangents and normals to functions: Rolle's Thorem and The Mean Value Theorem.

Before stating Rolle's Theorem, it is important to mention the *Extreme Value Theorem*. This theorem states that, if a function f is continuous on $[a, b]$, then f has both a minimum and a maximum on the interval.

Rolle's Theorem: If a function f is continuous on $[a, b]$ and differentiable on (a, b) and if $f(a) = f(b) = 0$, then there is at least one number c between a and b at which $f'(c) = 0$.

Mean Value Theorem: If a function f is continuous on $[a, b]$ and differentiable on (a, b) then there is at least one number c between a and b at which $f'(c) = \dfrac{f(b) - f(a)}{b - a}$.

A tangent line to a curve is found by taking the derivative, evaluating it at the appropriate value of x, which yields the slope of the tangent line, and then using the point slope form to find the equation of the line. The **normal** is a line that is perpendicular to the tangent at the point of tangency.

> **The normal is a line that is perpendicular to the tangent at the point of tangency.**

π TRY IT   Practice Activities

1. Given $f(x) = \ln(x-1)$,

    (a) Find the equation (in the $y = mx + b$ form) of the tangent to $f(x)$ at $x = 3$.

    (b) Find the zeroes of $f(x)$ analytically.

    (c) Sketch the graphs of $f(x)$ and the tangent line on the same axes.

    (d) Write your procedure for finding the zeroes of $f(x)$ graphically.

2. Let $f(x) = \dfrac{2}{9}x^{\frac{3}{2}}$ and suppose that the line $y = \dfrac{5}{2}x$ is parallel to the tangent of $f(x)$ at $x = a$. Find the value of a.

3. Let $f(x) = \sqrt{x-2}$ for $x \geq 2$. What are the possible x-values, usually denoted by c in the statement of the Mean Value Theorem, at which f' attains its mean value over the interval $2 \leq x \leq 5$?

4. If $H(x) = 2G(x) - \dfrac{1}{G(x)}$, $G(0) = 2$ and $G'(0) = -1$, find the exact value of $H'(0)$.

5. Let $f(x) = 2x^4 - 4x^2$. Find all c in the interval $(-3, 3)$ such that $f'(c) = 0$.

6. A tangent drawn to $y = -(x^2 - 4)$ at the point $(1, 3)$ forms a right triangle with the coordinate axes. Find the area of this right triangle.

7. Find the tangents to the curve $y = 2x^3 + 3x$ at the points where the slope is 9. What is the smallest slope on the curve? At what value of x does the curve have this slope?

8. Find the points on the curve $y = 4x^3 - 6x^2 - 24x + 30$ where the tangent is parallel to the x-axis.

9. Let $f(x) = 3^x - x^3$. Find the value(s) of x (correct to thousandths) such that the tangent to the curve is parallel to the secant through $(0, 1)$ and $(3, 0)$.

 **INSTRUCTION**

Optimization

> ## We need to have available tools to find minimum and maximum values.

One of the most common applications of calculus involves the determination of minimum and maximum values. Consider how frequently we hear or read terms like greatest profit, least cost, least time, optimum size, least area, greatest strength; we need to have available tools to find these values. To do these kinds of problems, one should follow this procedure:

1. It is usually very helpful to draw a figure and label it appropriately.

2. Write a **primary equation** for the quantity that is to be maximized (or minimized).

3. Reduce the primary equation to one having a single independent variable. This may involve the use of **secondary equations** relating the independent variables of the primary equation.

4. Determine the domain of the primary equation. That is, determine the values for which the stated problem makes sense. Always check the endpoints of the interval under consideration.

5. Use the techniques and concepts of differentiation, especially those for obtaining minimum and/or maximum values.

 **Practice Activities**

1. Find the coordinates of the point on the curve $y = \sqrt{x}$ closest to the point $(4, 0)$.

2. An indoor physical fitness room consists of a rectangular region with semicircles on each end. If the perimeter of the room is to be a 200-meter running track, find the dimensions that will make the area of the rectangular region as large as possible.

3. A circular cylinder container, open at the top and of capacity 24π cubic inches, is to be manufactured. If the cost of the material used for the bottom of the container is three times that used for the curved part, and if there is no waste of material, find the dimensions that will minimize the cost.

4. Find the dimensions of a right circular cylinder of maximum volume that can be inscribed in a sphere of radius 12.

5. A right triangle is formed in the first quadrant by the coordinate axes and a line segment passing through the point $(2, 3)$. Find the vertices of the triangle so that the area is a minimum.

6. What are the dimensions of the largest rectangle that can be inscribed in a right triangle of sides 5, 12, and 13 if one vertex of the rectangle is on the hypotenuse?

7. Two sand boxes are being constructed in a small playground. One is to have the shape of a regular hexagon, and the other that of an equilateral triangle. If 60 feet of board is available to make the sides, what is the length of a side of the hexagon if the total playing area is to be a minimum?

 **INSTRUCTION**

Related Rates

An important use of the Chain Rule is in related-rates problems. If several variables that are functions of time t are related by an equation, we can obtain a relation involving their (time) rates of change by differentiating with respect to t. The following is a suggested procedure for related-rates problems:

1. Assign symbols to all given quantities and quantities to be determined, and sketch and label the quantities appropriately.

2. Write an equation involving the variables for which rates of change either are given or are to be determined.

3. Using the Chain Rule, implicitly differentiate both sides of the equation with respect to time t.

4. Substitute into the resulting equation all known values for the variables and their rates of change.

5. Solve for the required rate of change.

40

Practice Activities

1. The surface area of a cube is increasing at the rate of 12 square inches per second. How fast is the volume of the cube increasing at the instant the surface area is 24 square inches?

2. The top of a silo has the shape of a hemisphere of diameter 20 feet. If it is coated uniformly with a layer of ice, and if the thickness is decreasing at a rate of 0.25 inches per hour, how fast is the volume changing when the ice is 2 inches thick?

3. A girl starts at a point A and runs east at a rate of 10 feet per second. One minute later another girl starts at point A and runs north at a rate of 8 feet per second. At what rate is the distance between them changing one minute after the second girl starts?

4. Let a parallelogram have sides of 8 and 12 and let vertex angle *A* be decreasing at a rate of 2° per minute. Find the rate of change of the area of the parallelogram when angle *A* equals 30°.

5. Ship A is sailing due south at 16 mph. At the same time, a second ship B, 32 miles south of A, is sailing due east at 12 mph. (a) At what rate are they approaching or separating at the end of one hour? (b) At what rate are they approaching or separating at the end of two hours? (c) When do they cease to approach each other and how far apart are they at that instant?

6. Water is running out of a conical container 12 feet in diameter and 8 feet deep (vertex down) and filling a spherical balloon. At the instant the depth of the water in the cone is 4 feet, the radius of the sphere is approximately 4 feet. The rate of change of the depth of the water in the cone at that instant is approximately _____ times the rate of change of the radius of the balloon.

INSTRUCTION

Finding Area Bounded by Functions

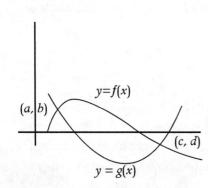

Using the figure on the left, the area bounded by $y = f(x)$, the x-axis, and the vertical lines $x = a$ and $x = b$ is found by subdividing the interval between a and b into n subintervals and the width of each subinterval is $\dfrac{b-a}{n} = \Delta x$. The area is found by adding rectangles with dimensions $f(c) \times \Delta x$, where c is some value of x between a and b.

By increasing the number of subintervals, the area found approaches the actual area. This process for finding area can be summarized below by the definite integral as the area of a region:

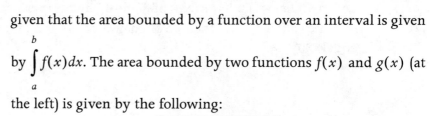

$$\text{Area} = \lim_{n \to \infty} \sum_{i=1}^{n} f(c_i)\Delta x = \int_{a}^{b} f(x)\,dx, \text{ if } f(x) \geq 0 \text{ for all } x \text{ on } [a, b],$$

$$\text{where } x_{i-1} \leq c_i \leq x_i \text{ and } \int_{a}^{b} f(x)\,dx = F(b) - F(a)$$

given that the area bounded by a function over an interval is given by $\displaystyle\int_{a}^{b} f(x)\,dx$. The area bounded by two functions $f(x)$ and $g(x)$ (at the left) is given by the following:

$$\int_{a}^{c} [f(x) - g(x)]\,dx$$

To find the area of the region between two intersecting graphs, you must find the points of intersection of the graphs. This can be done by either setting the two functions equal to each other and solving for x or by using a graphing calculator. Also, if two curves intersect at more than two points, then to find the area of the region between the curves, we must find all points of intersection and check to see which curve is above the other in each interval determined by these points.

π TRY IT   **Practice Activities**

1. Find the exact area of the region bounded by $x = 3 - y^2$ and $y = x - 1$.

2. Find the exact area bounded by $f(x) = 2 - x^2$ and $g(x) = x$ over $[-2, 2]$.

3. Find the exact area of the triangle with vertices $(2, -3)$, $(4, 6)$, and $(6, 1)$.

4. The region bounded by $y = \dfrac{1}{\sqrt{x}}$, $y = 0$, $x = 1$, and $x = 9$ is to be divided by two vertical lines, $x = a$ and $x = b$, into 3 regions equal in area; $1 < a < b < 9$. Find the value of a.

5. Find the area bounded by $y = x^{\frac{1}{m}}$ and $y = x^m$.

6. Find the area of the region bounded by the graphs of $f(x) = 10x + x^2 - 3x^3$ and $g(x) = 2x^2 - 4x$. Express the area correct to thousandths. Write the graphing calculator expression(s) used to find this area.

 **INSTRUCTION**

Numerical Integration

The Trapezoidal Rule approximates short stretches of the curve with line segments.

It is always possible to approximate the value of a definite integral by other means. Recall that the area bounded above by $y = f(x)$, below by the x-axis, and vertically by $x = a$ and $y = b$, and $f(x)$ is nonnegative, is given by $\int_a^b f(x)\,dx$. The value of the definite integral is then approximated by dividing the area into n strips, approximating the area of each strip by a rectangle or other geometric figure, then summing these values.

The first of these numerical integration methods is the Trapezoidal Rule. The Trapezoidal Rule approximates short stretches of the curve with line segments. To estimate the area under the curve, we add the areas of the trapezoids formed by joining the ends of these segments to the x-axis. The formula for finding the area of a trapezoid is $\frac{h}{2}(b_1 + b_2)$. The height of each trapezoid is the distance between each subinterval, namely $h = \frac{b-a}{n}$. To approximate $\int_a^b f(x)\,dx$, use the following, which is the Trapezoidal Rule: $T = \frac{h}{2}(y_0 + 2y_1 + 2y_2 + \ldots + 2y_{n-1} + y_n)$.

Simpson's Rule is based on approximating curves with parabolic arcs instead of line segments. To use Simpson's Rule, the number of subintervals n <u>must be even</u> and $h = \frac{b-a}{n}$. The formula is given by

$$S = \frac{h}{3}(y_0 + 4y_1 + 2y_2 + 4y_3 + \ldots + 2y_{n-2} + 4y_{n-1} + y_n)$$

Last, the average value of a function is given by $\frac{1}{b-a} \int_a^b f(x)\,dx$.

π **TRY IT**   **Practice Activities**

1. If $\int_a^b f(x)\,dx = 8$, $a = 2$, f is continuous, and the average value of f on $[a, b]$ is 4, find the value of b.

2. If $f(x) = \dfrac{1}{2} + \dfrac{1}{2}\cos(2x)$ over $0 \le x \le \pi$, find the average value of f over this interval.

3. The mathematical model to approximate the speed of traffic in miles per hour along Main Street is $S(t) = 5t^2 - 3t - 4$, where t is the number of hours past noon. Compute the average speed of the traffic between the hours of 2:00 P.M. and 4:00 P.M.

4. Compute both the Trapezoidal Rule and Simpson's Rule estimate of $\int_{-\pi}^{2\pi} \dfrac{\sin x}{x}\,dx$ for $n = 10$. Express both calculations correct to thousandths.

5. Correct to thousandths, compare the actual area bounded by $f(x) = \sqrt{4 + x^3}$ on the interval [0, 2] with the approximate area using the Trapezoidal Rule and Simpson's Rule taking $n = 4$.

6. A golf couse designer is preparing a sand trap for the proposed "signature hole" for a new golf country club. The designer would like to determine the cost for filling the trap with sand to a depth of 9 in. The indicated measurements in the diagram below occur every 12 ft. Sand costs $1.65 per cubic ft. What is the cost for filling the trap with sand?

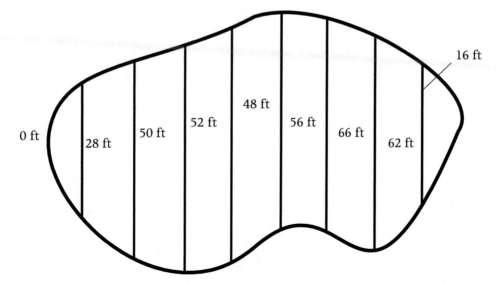

 **INSTRUCTION**

Volumes of Solids of Revolution

A **solid of revolution** is obtained when a plane region is revolved about a fixed line, called the axis of revolution. There are three major methods of obtaining the volume of a solid of revolution.

> **There are three major methods of obtaining the volume of a solid of revolution.**

Disc Method: We think of the rectangular strips of the bounded region being revolved about the *x*-axis; the solid of revolution is a disc (cylinder) with a height of Δx and a radius equal to the functional value. Because the formula for the volume of a disc (cylinder) is $\pi R^2 H$, the formula for finding the volume of the solid of revolution using discs when the region is revolved around the *x*-axis is

$$V = \pi \int_a^b [f(x)]^2 \, dx$$

If the region is to be revolved about the *y*-axis over the interval (c, d), the formula is:

$$V = \pi \int_c^d [g(y)]^2 \, dy$$

Washer Method: A washer is a disc with a hole in it. The volume may be regarded as the difference in the volumes of two concentric circles. The volume of a solid of revolution can be found viewing a typical washer as the difference of two discs. Thus, to find the volume of a region formed by two functions and two vertical strips $x = a$ and $x = b$ being revolved around the *x*-axis, we would use the formula

$$V = \pi \int_a^b ([f(x)]^2 - [g(x)]^2) \, dx$$

If the region is being revolved about the *y*-axis over the interval (c, d), the formula is

$$V = \pi \int_c^d ([f(x)]^2 - [g(y)]^2) \, dy$$

Shell Method: A cylindrical shell is a solid bounded by two concentric circular cylinders with the same height. If r_1 and r_2 are the radii of the inner and outer cylinders, respectively, and the common height is H, then the formula for the volume of the shell is

$$V = 2\pi \frac{r_1 + r_2}{2}(r_2 - r_1)H$$

With $r_2 - r_1 = T$, the thickness of the shell $\frac{r_1 + r_2}{2} = R$, the midradius, and H the height, the volume of the shell can now be written as

$$V = 2\pi R H T$$

The volume of a region bounded above by a function $f(x)$, below by the x-axis, and by the vertical lines $x = a$ and $y = b$ revolved around the y-axis is

$$V = 2\pi \int_a^b x f(x)\,dx$$

If the region is bounded above by $y = d$, below by $y = c$, to the right by $x = g(y)$, and to the left by the y-axis, the formula for finding the volume of the solid of revolution would be

$$V = 2\pi \int_a^d y g(y)\,dy$$

47

 **Practice Activities**

1. Oil in a spherical tank 40 ft in diameter is 15 ft deep. How much oil is contained in the tank? Express your answer in terms of π.

2. Find the volume of the solid generated by revolving around the x-axis the region bounded by $y = x^2$ and $y = 2 - x^2$.

3. In the first quadrant, the line $y = 3x$ divides the area bounded by $y = 4 - x^2$ into two distinct regions: Region A, which lies above the line, and Region B, which lies below the line. If these areas were revolved around the x-axis, which would yield the largest volume?

4. Using two distinct methods, find the volume of the solid generated when the area bounded by $y = x^2$ and $x = y^2$ is revolved around the y-axis.

5. A nose cone for a space reentry vehicle is designed so that a cross section, taken x ft from the tip and perpendicular to the axis of symmetry, is a circle of radius $0.25x^2$ ft. Find the volume (correct to hundredths) of the nose cone given that its length is 20 ft.

6. Let $y = \cos(2x^2)$ for x in the interval $[0, c]$, where c is the first positive x-intercept. Find the volume of the region, expressed as an exact value, between $y = f(x)$ and the x-axis containing the interval $[0, c]$, rotating about the y-axis. Write, in simplest form, the definite integral expression for the volume of the region between $y = 1$ and $y = f(x)$ on the given interval, rotating about the x-axis. Write an expression for finding the volume using a graphing calculator. Find the volume correct to thousandths.

 **INSTRUCTION**

Areas of Surfaces of Revolution

As previously stated in "Volumes of Solids of Revolution," a solid of revolution is obtained when a plane region is revolved about a fixed line, called the axis of revolution. The summation character of the definite integral is used to find the arc length of a plane curve and the area of a surface of revolution. If the function $y = f(x)$ represents a smooth curve on the interval $[a, b]$, then the arc length of f between a and b is given by

$$L = \int_a^b \sqrt{1 + [f'(x)]^2}\, dx$$

Similarly, for a smooth curve expressed in the form $x = g(y)$ between c and d, the length of the arc is given by

$$L = \int_c^d \sqrt{1 + [g'(y)]^2}\, dy$$

If the curve is represented parametrically with $x = x(t)$ and $y = y(t)$ over the interval $[a, b]$, the length of the arc is given by

$$L = \int_a^b \sqrt{\left(\frac{dx}{dt}\right)^2 + \left(\frac{dy}{dt}\right)^2}\, dt$$

For areas of surfaces of revolution, the following summarizes the formulas to be used.

Axis of Revolution	Formula
x-axis	$S = 2\pi \int_a^b f(x) \sqrt{1 + [f'(x)]^2}\, dx$
y-axis	$S = 2\pi \int_a^b x \sqrt{1 + [f'(x)]^2}\, dx$

In general, if $y = f(x)$ has a continuous derivative on $[a, b]$, the area S of the surface of revolution formed by revolving the graph of f about a horizontal axis is

$$S = 2\pi \int_a^b r(x) \sqrt{1 + [f'(x)]^2}\, dx$$

where $r(x)$ is the distance between the graph and the axis of revolution.

For functions expressed parametrically, where $x = x(t)$ and $y = y(t)$ over the interval $[a, b]$,

Axis of Revolution	Formula
x-axis	$S = 2\pi \int_a^b y(t) \sqrt{\left(\dfrac{dx}{dt}\right)^2 + \left(\dfrac{dy}{dt}\right)^2}\, dt$
y-axis	$S = 2\pi \int_a^b x(t) \sqrt{\left(\dfrac{dx}{dt}\right)^2 + \left(\dfrac{dy}{dt}\right)^2}\, dt$

π **TRY IT**   **Practice Activities**

1. Find the length of the arc, correct to thousandths, of the curve $x = t^2$ and $y = t^3$ between $t = 0$ and $t = 2$.

2. Find the exact value for the surface area of a solid generated by revolving about the y-axis the area in the first quadrant enclosed by the functions $f(x) = x^2$ and $g(x) = 2 - x^2$.

3. Find the surface area of a zone of a sphere obtained by revolving about the x-axis the part of the semicircle $y = \sqrt{a^2 - x^2}$, $-a < b < c < a$.

4. Express the surface area of the solid generated by revolving the area bounded by $y = 9 - x^2$, $y = 0$, $x = 2$, $x = 0$ about the y-axis as an exact value.

 **INSTRUCTION**

Work

The concept of work is important to scientists and engineers for determining the energy needed to perform various tasks. For instance, it is useful to know the amount of work done when a crane lifts a steel girder, when a spring is compressed, when a rocket is propelled into the air, or when a truck pulls a load along a highway. In general, we say that **work** is done by a force when it moves an object. If the force applied to the object is constant, we have the following definition of work:

> In general, we say that work is done by a force when it moves an object.

If an object is moved a distance D in the direction of an applied constant force F, then the work W done by the force is defined as $W = FD$.

If a variable force is applied to an object, then calculus is needed to determine the work done because the amount of force changes as the object changes positions. This leads to the definition of work done by a variable force:

If an object is moved along a straight line by a continuously varying force $F(x)$, then the work W done by the force as the object is moved from $x = a$ to $x = b$ is given by

$$W = \int_a^b F(x)\,dx$$

 **Practice Activities**

1. A spring exerts a force of 100 N when it is stretched 0.2 meters beyond its natural length. How much work is required to stretch the spring 0.8 meters beyond its natural length?

2. A cylinderical tank of radius 5 ft and height 9 ft is two thirds filled with water. Find the work required to pump all the water over the upper rim.

3. A rocket weighing 3 tons is filled with 40 tons of liquid fuel. In the initial part of the flight, fuel is burned off at a constant rate of 2 tons per 1,000 ft of vertical height. How much work is done in lifting the rocket to 3,000 ft?

 **INSTRUCTION** ## Techniques of Integration

In this part, we will work with integrals for which antiderivatives are not readily recognized but can be put into a form that will allow for paper-and-pencil solutions. A major step in solving any integration problem is recognizing the proper basic integration formula to be used. One of the challenges about integration is the fact that slight differences in the integrand can lead to very different solution techniques. Some techniques that will be helpful are as follows: using two basic formulas to solve a single integral, breaking a quotient into two parts, knowing the integral may be a disguised form of a particular rule, using algebraic procedures, and using trigonometric identities.

The following is a list of integral formulas from previous parts that will be helpful and a few more additional ones:

$$\int kf(x)dx = k\int f(x)dx, \quad \int [f(x)+g(x)]dx = \int f(x)dx + \int g(x)dx, \quad \int u^n du = \frac{u^{n+1}}{n+1} + C$$

$$\int \cos u\, du = \sin u + C \qquad\qquad \int \csc^2 u\, du = -\cot u + C$$

$$\int \sin u\, du = -\cos u + C \qquad\qquad \int \sec u \tan u\, du = \sec u + C$$

$$\int \tan u\, du = \ln|\sec u| + C \text{ or } -\ln|\cos u| + C \quad \int \csc u \cot u\, du = -\csc u + C$$

$$\int \cot u\, du = -\ln|\csc u| + C \text{ or } \ln|\sin u| + C \quad \int \sec u\, du = \ln|\sec u + \tan u| + C$$

$$\int \sec^2 u\, du = \tan u + C \qquad\qquad \int \csc u\, du = \ln|\csc u - \cot u| + C$$

$$\int e^u du = e^u + C \qquad\qquad\qquad \int a^u du = \frac{a^u}{\ln a} + C (a > 0, a \neq 1)$$

$$\int \frac{du}{u} = \ln|u| + C$$

54

$$\int \frac{du}{\sqrt{1-u^2}} = \sin^{-1} u + C \qquad\qquad \int \frac{du}{1+u^2} = \tan^{-1} u$$

$$\int \frac{du}{u\sqrt{u^2-1}} = \int \frac{d(-u)}{(-u)\sqrt{u^2-1}} = \sec^{-1}|u| + C = \cos^{-1}\left|\frac{1}{u}\right| + C$$

$$\int \frac{du}{a^2+u^2} = \frac{1}{a}\tan^{-1} u + C \qquad\qquad \int \frac{du}{\sqrt{a^2-u^2}} = \sin^{-1}\frac{u}{a} + C$$

$$\int \frac{du}{\sqrt{a^2+u^2}} = \int \sec\theta\, d\theta = \ln(|\sec\theta + \tan\theta| + C) = \ln\left|\sqrt{a^2+u^2} + u\right| + C$$

$$\int \frac{du}{\sqrt{u^2-a^2}} = \ln\left|u + \sqrt{u^2-a^2}\right| + C$$

π   TRY IT   **Practice Activities**

1. Evaluate $\displaystyle\int \frac{dx}{2x^2 - 8x + 10}$.

2. Without the use of a graphing calculator, find $\displaystyle\int_{0}^{\frac{\pi}{4}} (\sec^4 x \tan x)\, dx$.

3. If $\displaystyle\int \frac{F\,dx}{ax^2 + bx + c} = \ln|ax^2 + bx + c| + C$, find F.

4. Evaluate the following integrals:

   (a) $\displaystyle\int \frac{x^2\,dx}{\sqrt{16 - x^6}}$

   (b) $\displaystyle\int \cot x (\ln(\sin x))\, dx$

   (c) $\displaystyle\int (\tan\theta + \sec\theta)^2\, d\theta$

   (d) $\displaystyle\int \frac{dx}{1 + \sin x}$

5. Express $\displaystyle\int_{1}^{2} \frac{2^{\frac{1}{x}}}{x^2}\, dx$ as an exact value.

Answer Key

Limits and Continuity

1. Substituting $x = 2$ into $\dfrac{3}{x+1}$ yields a limit, $L = 1$. The procedure for the proof involves that, given any radius $\varepsilon > 0$ about L, there exists a radius $\delta > 0$ about x_0 such that, for all x, $0 < |x - x_0| < \delta$ implies that $|f(x) - L| < \varepsilon$.

   (a) $\left| \dfrac{3}{x+1} - L \right| < \varepsilon$

   $-\varepsilon < \dfrac{3}{x+1} - 1 < \varepsilon$

   $1 - \varepsilon < \dfrac{3}{x+1} < 1 + \varepsilon$

   $\dfrac{1}{1-\varepsilon} > \dfrac{x+1}{3} > \dfrac{1}{1+\varepsilon}$

   $\dfrac{3}{1-\varepsilon} > \dfrac{x+1}{1} > \dfrac{3}{1+\varepsilon}$

   $\dfrac{3}{1-\varepsilon} - 3 > x + 1 - 3 > \dfrac{3}{1+\varepsilon} - 3$

   $\dfrac{3\varepsilon}{1-\varepsilon} > x - 2 > \dfrac{-3\varepsilon}{1+\varepsilon}$

   Choose min $\left\{ \dfrac{3\varepsilon}{1+\varepsilon}, \dfrac{3\varepsilon}{1-\varepsilon} \right\}$, which is

   $\dfrac{3\varepsilon}{1+\varepsilon} = \delta \ \therefore 0 < |x - 2| < \dfrac{3\varepsilon}{1+\varepsilon}.$

   (b) If $\varepsilon = 0.005$, then $\delta = \dfrac{3(0.005)}{1 + 0.005} = 0.015$, then $(2 - \delta, 2 + \delta) = (1.985, 2.015)$.

2. (a) $\displaystyle\lim_{x \to 0} \dfrac{\dfrac{1}{2+x} - \dfrac{1}{2}}{x}$

   $\displaystyle\lim_{x \to 0} \dfrac{2 - 2 - x}{2x(2+x)}$

   $\displaystyle\lim_{x \to 0} \dfrac{-1}{2(2+x)} = \dfrac{-1}{4}$

(b) $\lim\limits_{x \to 2} \dfrac{x^3 - 8}{x^2 - 4}$

$\lim\limits_{x \to 2} \dfrac{(x - 2)(x^2 + 2x + 4)}{(x - 2)(x + 2)}$

$\lim\limits_{x \to 2} \dfrac{(x^2 + 2x + 4)}{(x + 2)} = \dfrac{4 + 4 + 4}{4} = 3$

(c) $\lim\limits_{x \to 0} \dfrac{(\csc 5x)(\sec 8x)}{(\csc 8x)(\sec 3x)}$

$\lim\limits_{x \to 0} \dfrac{1}{\sin 5x} \times \dfrac{1}{\cos 8x} \times \dfrac{\sin 8x}{1} \times \dfrac{\cos 3x}{1}$

$\lim\limits_{x \to 0} \dfrac{1}{\sin 5x} \times \dfrac{1}{\cos 8x} \times \dfrac{\sin 8x}{1} \times \dfrac{\cos 3x}{1} \times \dfrac{5x}{8x} \times \dfrac{8}{5} =$

$\lim\limits_{x \to 0} \dfrac{5x}{\sin 5x} \times \dfrac{1}{\cos 8x} \times \dfrac{\sin 8x}{8x} \times \dfrac{\cos 3x}{1} \times \dfrac{8}{5}$

$1 \times 1 \times 1 \times 1 \times \dfrac{8}{5} = \dfrac{8}{5}$

(d) $\lim\limits_{x \to 0} \dfrac{x + \sin 2x}{x}$

$\lim\limits_{x \to 0} \dfrac{x}{x} + \dfrac{2 \sin x \cos x}{x}$

$\lim\limits_{x \to 0} \left(\dfrac{x}{x} + \dfrac{2}{1} \times \dfrac{\sin x}{x} \times \dfrac{\cos x}{1} \right)$

$1 + (2)(1)(1) = 3$

(e) $\lim\limits_{x \to 0} \dfrac{\cos 2x}{\sin x - \cos x}$

$\lim\limits_{x \to 0} \dfrac{\cos^2 x - \sin^2 x}{\sin x - \cos x}$

$\lim\limits_{x \to 0} \dfrac{(\cos x - \sin x)(\cos(x + \sin x))}{\sin x - \cos x}$

$\lim\limits_{x \to 0} (-\cos x - \sin x) = -1$

3. The function is discontinuous at $x = -2$ because the function is undefined at $x = -2$.

The function is discontinuous at $x = 0$ because $f(0) = 2$ and $\lim\limits_{x \to 0^-} f(x) = -1$ *and*

$\lim\limits_{x \to 0^+} f(x) = 0$ and $\lim\limits_{x \to 0^-} f(x) = 1 \neq \lim\limits_{x \to 0^+} f(x) = 0 \neq f(0)$.

The function is discontinuous at $x = 2$ because

$f(2) = 0$ and $\lim\limits_{x \to 2^-} f(x) = \lim\limits_{x \to 2^1} f(x) = -2$ and

$f(2) \neq \lim\limits_{x \to 2^-} f(x) = \lim\limits_{x \to 2^+} f(x) = 2.$

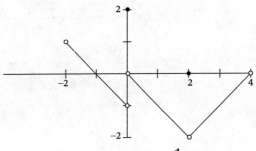

4. $\lim\limits_{x \to 3^-} f(x) = \lim\limits_{x \to 3^-} (x^2 - 1) = 8;$

also, $\lim\limits_{x \to 3^-} f(x) = \lim\limits_{x \to 3^+} 2bx = 2b(3) = 6b;$

thus, $6b = 8$ and $b = \dfrac{4}{3}.$

The graph of $\begin{cases} x^2 - 1, \, x < 3 \\ \dfrac{8}{3}x, \quad x \geq 3 \end{cases}$

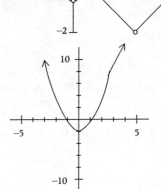

5. At $x = 2$, $f(x) = 8$. For $f(x)$ to be continuous at $x = 2$, $2^3 = 2(2) + A$ and $A = 4.$

6. (a) $\lim\limits_{x \to \infty} \dfrac{3 - x}{4 + x + x^2} = \lim\limits_{x \to \infty} \dfrac{\dfrac{3}{x^2} - \dfrac{x}{x^2}}{\dfrac{4}{x^2} + \dfrac{x}{x^2} + \dfrac{x^2}{x^2}} = \lim\limits_{x \to \infty} \dfrac{\dfrac{3}{x^2} - \dfrac{1}{x}}{\dfrac{4}{x^2} + \dfrac{1}{x} + 1} = \dfrac{0 + 0}{0 + 0 + 1} = 0.$

(b) $\lim\limits_{x \to \infty} \dfrac{4x^4 + 5x + 1}{37x^3 - 9} = \lim\limits_{x \to \infty} \dfrac{\dfrac{4x^4}{x^4} + \dfrac{5x}{x^4} + \dfrac{1}{x^4}}{\dfrac{37x^3}{x^4} - \dfrac{9}{x^4}} = \lim\limits_{x \to \infty} \dfrac{4 + \dfrac{5}{x^3} + \dfrac{1}{x^4}}{\dfrac{37}{x} - \dfrac{9}{x^4}} = \infty \text{ (no limit).}$

(c) $\lim\limits_{x \to \infty} \dfrac{x^3 - 4x^2 + 7}{3 - 6x - 2x^3} = \lim\limits_{x \to \infty} \dfrac{\dfrac{x^3}{x^3} + \dfrac{4x^2}{x^3} + \dfrac{7}{x^3}}{\dfrac{3}{x^3} - \dfrac{6x}{x^3} - \dfrac{2x^3}{x^3}} = \lim\limits_{x \to \infty} \dfrac{1 - \dfrac{4}{x} + \dfrac{7}{x^3}}{\dfrac{3}{x^3} - \dfrac{6}{x^2} - 2} = \dfrac{1 - 0 + 0}{0 - 0 - 2} = -\dfrac{1}{2}.$

The Derivative

1. (a) Using the definition $f'(x) = \lim\limits_{h \to 0} \dfrac{f(x+h) - f(x)}{h},$

$f'(x) = \lim\limits_{h \to 0} \dfrac{\dfrac{2(x+h) - 1}{(x+h) + 2} - \dfrac{2x - 1}{x + 2}}{h}$

$$f'(x) = \lim_{h \to 0} \frac{(x+2)(2x+2h-1)-(2x-1)(x+h+2)}{(x+2)(x+h+2)(h)}$$

$$f'(x) = \lim_{h \to 0} \frac{2x^2+2xh+3x+4h-2-2x^2-2xh-3x+h+2}{(x+2)(x+h+2)(h)}$$

$$f'(x) = \lim_{h \to 0} \frac{5h}{(x+2)(x+h+2)(h)}$$

$$f'(x) = \lim_{h \to 0} \frac{5}{(x+2)^2}$$

(b) Using the definition $f'(x) = \lim\limits_{h \to 0} \dfrac{f(x+h)-f(x)}{h}$,

$$f'(x) = \lim_{h \to 0} \frac{\sqrt{3(x+h)-1}-\sqrt{3x-1}}{h}$$

$$f'(x) = \lim_{h \to 0} \frac{\sqrt{3(x+h)-1}-\sqrt{3x-1}}{h} \times \frac{\sqrt{3(x+h)-1}+\sqrt{3x-1}}{\sqrt{3(x+h)-1}+\sqrt{3x-1}}$$

$$f'(x) = \lim_{h \to 0} \frac{3(x+h)-1-(3x-1)}{(\sqrt{3(x+h-1)}+\sqrt{3x-1})(h)}$$

$$f'(x) = \lim_{h \to 0} \frac{3h}{(\sqrt{3(x+h-1)}+\sqrt{3x-1})(h)}$$

$$f'(x) = \frac{3}{2\sqrt{3x-1}}$$

(c) $f(x) = 3x^4$. By definition, $f'(x) = \lim\limits_{h \to 0} \dfrac{f(x+h)-f(x)}{h}$,

$$f'(x) = \lim_{h \to 0} \frac{3(x+h)^4-3x^4}{h}.$$

Expand $(x+h)^4$ using Pascal's Triangle,

$$f'(x) = \lim_{h \to 0} \frac{3x^4+12x^3h+18x^2h^2+12xh^3+h^4-3x^4}{h}$$

$$f'(x) = \lim_{h \to 0} \frac{12x^3h+18x^2h^2+12xh^3+h^4}{h}$$

$$f'(x) = \lim_{h \to 0} (12x^3+18x^2h+12xh^2+h^3)$$

$$f'(x) = 12x^3$$

Numerical Derivatives

1. Using a graphing calculator, let $Y_1 = 4^x$. nDeriv(Y_1, x, 0) = 1.386294805.

   Using a graphing calculator, $\dfrac{4^{0.08} - 4^{-0.08}}{(2)(0.08)} = 1.38913792$.

   Therefore, the symmetric difference is accurate to two places.

2. Using a graphing calculator and the symmetric difference quotient for $f(x) = x^4$ to find $f'(1)$ and $h = 0.01$,

   $$f'(1) = \frac{f(1+0.01) - f(1-0.01)}{2(.01)} = \frac{(1.01)^4 - (.99)^4}{.02} = 4.0004.$$

   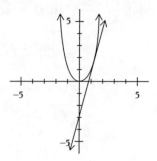

   NDER $(f(x), x, 1) \Rightarrow 4.000004$. The exact value is 4.

   Using NDER $(f(x), x, 1)$ as the slope and $(1, 1)$, the equation of the line is $y - 1 = 4.000004(x - 1)$, $y = 4.000004x - 3.000004$, or $y = 4x - 3$.

Rules for Differentiation

1. (a) Rewrite $y = (x^2 + 2x - 1)^{\frac{1}{2}}$

   $$y' = \frac{1}{2}(x^2 + 2x - 1)^{\frac{-1}{2}}(2x + 2)$$

   $$y' = \frac{2(x+1)}{2\sqrt{x^2 + 2x - 1}} = \frac{x+1}{\sqrt{x^2 + 2x - 1}}$$

   (b) Use the quotient rule,

   $$y' = \frac{\sqrt{1 - x^2}(1) - x\dfrac{1(-2x)}{2\sqrt{1 - x^2}}}{1 - x^2}$$

   $$y' = \frac{\dfrac{1 - x^2 + x^2}{\sqrt{1 - x^2}}}{1 - x^2} = \frac{1}{(1 - x^2)^{\frac{3}{2}}}$$

   (c) Use the formula $\dfrac{d}{dx}\tan^{-1} u = \dfrac{1}{1 + u^2}\dfrac{du}{dx}$,

   $$y' = \frac{\dfrac{1}{2}}{1 + \dfrac{x^2}{4}} = \frac{2}{4 + x^2}$$

(d) Use the following formulas:

$$\frac{d}{dx}(\ln u) = \frac{1}{u}\frac{du}{dx}$$

$$\frac{d}{dx}(\sec u) = \sec u \tan u \frac{du}{dx}$$

$$\frac{d}{dx}(\tan u) = \sec^2 u \frac{du}{dx}$$

$$y' = \frac{\sec x \tan x + \sec^2 x}{\sec x + \tan x} = \sec x$$

(e) Use the Quotient Rule and

$$\frac{d}{dx}(e^u) = e^u \frac{du}{dx}$$

$$y' = \frac{(e^x + e^{-x})(e^x + e^{-x}) - (e^x - e^{-x})(e^x - e^{-x})}{(e^x + e^{-x})^2}$$

$$y' = \frac{(e^{2x} + 2 + e^{-2x}) - (e^{2x} - 2 + e^{-2x})}{(e^x + e^{-x})^2}$$

$$y' = \frac{4}{(e^x + e^{-x})^2}$$

2. (a) Use the Product Rule and

$$\frac{d}{dx}\sin u = \cos u \frac{du}{dx}$$

$$\frac{dy}{dx} = x^2 \cos\frac{1}{x}\left(-\frac{1}{x^2}\right) + \left(\sin\frac{1}{x}\right)(2x)$$

$$\frac{dy}{dx} = 2x\sin\frac{1}{x} - \cos\frac{1}{x}$$

(b) Use a trig ID and

$$\frac{d}{dx}\csc u = -\csc u \cot u \frac{du}{dx}$$

Because $y = \frac{1}{2}\csc 2x$,

$$\frac{dy}{dx} = \frac{1}{2}(-2\csc 2x \cot 2x)$$

$$\frac{dy}{dx} = -\csc 2x \cot 2x$$

(c) Use the Power Rule and

$$\frac{d}{dx}\sec u = \sec u \tan u \frac{du}{dx}$$

$$\frac{dy}{dx} = 2\sec\sqrt{x}\,\sec\sqrt{x}\,\tan\sqrt{x}\,\frac{1}{2\sqrt{x}}$$

$$\frac{dy}{dx} = \frac{\sec^2\sqrt{x}\,\tan\sqrt{x}}{\sqrt{x}}$$

(d) Use the Product Rule and the Power Rule and

$$\frac{d}{dx}(\ln u) = \frac{1}{u}\frac{du}{dx}$$

$$\frac{dy}{dx} = \frac{x(3\ln^2 x)}{x} + \ln^3 x$$

$$\frac{dy}{dx} = 3\ln^2 x + \ln^3 x$$

(e) Use the Power Rule and

$$\frac{d}{dx}(\sin^{-1} u) = \frac{1}{\sqrt{1-u^2}}\frac{du}{dx}$$

$$\frac{dy}{dx} = \frac{1}{\sqrt{1-x^2}} - \frac{1(-2x)}{2\sqrt{1-x^2}}$$

$$\frac{dy}{dx} = \frac{1+x}{\sqrt{1-x^2}}$$

3. $f'(x) = 100(2x-1)^{99}(2^1)$

$$f''(x) = (100)(99)(2x-1)^{98}(2^2) = \frac{100!}{98!}(2x-1)^{98}(2^2)$$

$$f'''(x) = (100)(99)(98)(2x-1)^{97}(2^3) = \frac{100!}{97!}(2x-1)^{97}(2^3)$$

The pattern continues: $f^{10}(x) = \frac{100!}{90!}(2x-1)^{90}(2^{10})$;

therefore, $a = 100, b = 90, c = 10, d = 90$.

4. $$\frac{dy}{d\left(\frac{1}{1-x}\right)} = \frac{\frac{dy}{dx}}{\frac{d\left(\frac{1}{1-x}\right)}{dx}} = \frac{2x+1}{\frac{1}{(1-x)^2}} = (2x+1)(1-x)^2$$

Velocity, Speed, and Other Rates of Change

1. (a) Given $C(x) = 2x^3 - 5x^2 + 8$ $x \geq 2$, then the marginal cost equation is given by:

   $MC = C'(x) = 6x^2 - 10x$.

   (b) $C'(10) = 6(10)^2 - 10(10) = 500$.

   (c) The interpretation of part (b) is that the marginal cost of the 11th unit is $500.

2. (a) Given $C(x) = 20 + 2x + 0.01x^2$, $x \geq 0$, then the marginal cost equation is given by:
    $MC = C'(x) = 2 + 0.02x$.

    (b) $C'(100) = 2 + 0.02(100) = \$4$ and $C'(200) = 2 + 0.02(200) = \$6$.

    (c) At a production level of $x = 100$, one more unit costs approximately \$4 to produce. At a production level of $x = 200$, one more unit costs approximately \$6 to produce.

    (d) $C(100) = 20 + 2(100) + 0.01(100)^2 = \$320$.

    (e) Thus, the cost of producing the 101st unit is $C(101) - C(100) = \$4.01$, or the cost of producing the 101st unit is \$4.01.

3. (a) Given $H(x) = -16t^2 + 64t + 80$, then the instantaneous velocity is $H'(x) = -32t + 64$ and $H'(1) = 32$ ft/sec.

    (b) $H(4) = -16(4)^2 + 64(4) + 80 = 80$. The ball will be 80 ft above the ground after 4 sec.

    (c) $H'(4) = -32(4) + 64 = -64$ ft/sec. The negative denotes that the ball is falling at the rate of 64 ft/sec.

    (d) Initial velocity occurs when $t = 0$. $H'(0) = -32(0) + 64 = 64$ ft/sec.

    (e) The maximum height occurs when the velocity $H'(x) = 0$. Thus, when $-32t + 64 = 0$, $t = 2$. $H(2) = -16(2)^2 + 64(2) + 80 = 144$ ft.

    (f) The ball hits the ground when $H(t) = 0$. Solve $-16t^2 + 64t + 80 = 0$, which yields $t = 5$ and $t = -1$. Discard $t = -1$ and the ball hits the ground in 5 sec.

    (g) The velocity of the ball hitting the ground at $t = 5$ is $H'(5) = -32(5) + 64 = -96$ ft/sec. minus ninety-six (–96) denotes that the ball is moving downward and hits the ground at 96 ft/sec.

4. (a) Given $N(t) = \frac{1}{3}t^3 - 6t^2 + t + 1{,}757$, then $N'(t) = t^2 - 12t + 1$.

    (b) $N'(6) = (6)^2 - 12(6) + 1 = -35$, which denotes that sales were decreasing at that point in time.

    (c) $N'(12) = (12)^2 - 12(12) + 1 = 1$, which denotes that sales were increasing at that point in time.

5. Given $Q(t) = 200(30 - t)^2 = 200(900 - 60t + t^2)$, $Q'(t) = 200(-60 + 2t) = 400(t - 30)$. $Q'(10) = 400(10 - 30) = -8{,}000$ gal/min. At the end of 10 min, water is flowing out at the rate of 8,000 gal/min. During the 10-min interval, the average rate is found by

    $$\text{average rate } \frac{Q(10) - Q(0)}{10 - 0} = \frac{80{,}000 - 180{,}000}{10} = -10{,}000 \text{ gal/min.}$$

Interpreting Graphs of Functions and Their Derivatives

1. For $f(x) = \dfrac{x^4 + 1}{x^2}$,

   $f'(x) = \dfrac{2(x^4 - 1)}{x^3} = \dfrac{2(x^2 + 1)(x - 1)(x + 1)}{x^3}$;

   $f''(x) = \dfrac{2(x^4 + 3)}{x^4}$; $f'(x) = 0$ yields $x = 1$ and $x = -1$.

   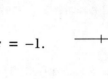

   The function is discontinuous at $x = 0$

   $f(-1) = f(1) = 2$. $f''(-1) = f''(1) > 0$; therefore,

   $(-1, 2)$ and $(1, 2)$ are relative minima.

   The function is decreasing on the intervals $(-\infty, -1)$ and

   $(0, 1)$.

   The function is increasing on the intervals $(-1, 0)$ and $(1, \infty)$.

   There is no point of inflection because $f''(x) > 0$ for all x, except at $x = 0$.

   The curve is concave upward for all values of x.

2. $f'(x) = 3x^2 - 10x + 3$ and $f''(x) = 6x - 10$.

   $f'(x) = 0$ at $x = 3$ and $x = \dfrac{1}{3}$.

   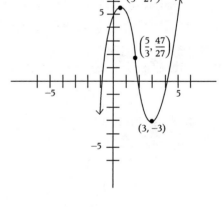

   Because $f''(3) > 0$, then $(3, f(3))$ or $(3, -3)$ is a local minimum.

   Because $f''\left(\dfrac{1}{3}\right) < 0$, then $\left(\dfrac{1}{3}, f\left(\dfrac{1}{3}\right)\right) = \left(\dfrac{1}{3}, \dfrac{175}{27}\right)$ is a local maximum.

   $f''(x) = 0$ at $x = \dfrac{5}{3}$, then $\left(\dfrac{5}{3}, \dfrac{47}{27}\right)$ is a point of inflection.

   The function is increasing over the intervals $\left(-\infty, \dfrac{1}{3}\right)$ and

   $(3, \infty)$ because $f'(x) > 0$ over these intervals. The function is decreasing over $\left(\dfrac{1}{3}, 3\right)$ because

   $f'(x) > 0$ in this interval.

   Because $f''(x) < 0$ over $\left(-\infty, \dfrac{5}{3}\right)$, the function is concave down. The function is concave up

   over $\left(\dfrac{5}{3}, \infty\right)$ because $f''(x) > 0$.

3. $f'(x) = \dfrac{-10x}{(x^2 - 4)^2}$ and $f''(x) = \dfrac{10(3x^2 + 4)}{(x^2 - 4)^3}$.

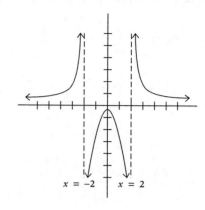

The function is discontinuous at $x = \pm 2$, so there are vertical asymptotes at these values.

$f'(x) = 0$ at $x = 0$. $\left(0, \dfrac{-1}{4}\right)$ is a relative maximum

because $f''(0) < 0$.

On the intervals $(-\infty, -2)$ and $(-2, 0)$, $f'(x) > 0$, so the function is increasing on these intervals.

On the intervals $(0, 2)$ and $(2, \infty)$, $f'(x) < 0$, so the function is decreasing on these intervals.

There are no points where $f''(x) = 0$, so there is no point of inflection.

For the intervals $(-\infty, -2)$ and $(2, \infty)$, $f''(x) > 0$, so the curve is concave upward.

For the interval $(-2, 2)$, $f'(x) > 0$, so the curve is concave downward.

4. $f'(x) = \dfrac{3}{(x - 1)^2}$ and $f''(x) = \dfrac{6}{(x - 1)^3}$.

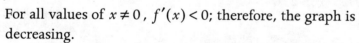

The function is discontinuous at $x = 1$, so there is a vertical asymptote at this value of x.

There are no values of x for which $f'(x)$ and $f''(x)$ equal zero.

$\lim\limits_{x \to \pm\infty} f(x) = 2$; therefore, there is a horizontal asymptote at

$y = 2$.

For all values of $x \neq 0$, $f'(x) < 0$; therefore, the graph is decreasing.

For the interval $(-\infty, 1)$, the graph is concave downward because $f''(x) < 0$.

For the interval $(1, \infty)$, the graph is concave upward because $f''(x) > 0$.

5. $f'(x) = 3x^2 + 2ax + b$ and $f''(x) = 6x + 2a$ and, because f has a relative maximum at $x = 1$,

then $f'(1) = 3(1)^2 + 2a(1) + b = 0 \Rightarrow 2a + b + 3 = 0$.

Because f has an inflection point at $x = 2$, $f''(2) = 0 \Rightarrow f''(2) = 6(2) + 2a \Rightarrow a = -6$; also

$f''(1) < 0$, which verifies a relative maximum at $x = 1$.

To find b, $2(-6) + b + 3 = 0$ and $b = 9$

To find c, $f(0) = 6 = (0)^3 + 6(0)^2 + 9(0) + c$ and $c = 6$.

Derivatives and Integrals of Trigonometric Functions

1. (a) $\dfrac{dy}{dx} = 3\cos^2 x - 3\sin^2 x = 3(\cos^2 x - \sin^2 x) = 3\cos(2x)$

   $y = \displaystyle\int 3\cos(2x)\,dx = \dfrac{3}{2}\int \cos u\,du$, where $u = 2x$ and $dx = \dfrac{du}{2}$

   $y = \dfrac{3}{2}\sin(2x) + C$

   (b) $\dfrac{dy}{dx} = x\cot(3x^2)$; let $u = 3x^2$, then $x\,dx = \dfrac{du}{6}$

   $y = \displaystyle\int x\cot(3x^2)\,dx = \dfrac{1}{6}\int \cot u\,du = \dfrac{1}{6}\ln|\sin u| + C = \dfrac{1}{6}\ln|\sin(3x^2)| + C$

2. $y = f(x) = \displaystyle\int \sin(\pi x)\,dx = -\dfrac{1}{\pi}\cos(\pi x) + C$

   because $y = f(x)$ and $f(0) = 0$, then $y = 0$. $0 = -\dfrac{1}{\pi}\cos(0) + C \Rightarrow C = \dfrac{1}{\pi}$.

   $f(x) = -\dfrac{1}{\pi}\cos(\pi x) + \dfrac{1}{\pi}$

   $f(-1) = -\dfrac{1}{\pi}\cos(-\pi) + \dfrac{1}{\pi} = -\dfrac{1}{\pi}(-1) + \dfrac{1}{\pi} = \dfrac{2}{\pi}$

3. $y = \csc(t + \sqrt{t}) = \dfrac{1}{\sin(t + \sqrt{t})} = \left(\sin\left(t + t^{\frac{1}{2}}\right)\right)^{-1}$

   $y' = -1\left(\sin\left(t + t^{\frac{1}{2}}\right)\right)^{-2}\cos\left(t + t^{\frac{1}{2}}\right)\left(1 + \dfrac{1}{2}t^{-\frac{1}{2}}\right)$

   $y'(1) = -1(\sin 2)^{-2}\cos 2\left(1 + \dfrac{1}{2}\right) = -\dfrac{3}{2}\dfrac{\cos 2}{\sin^2 2} = 0.755$

4. (a) $\displaystyle\int (\tan\theta - 1)^2\,d\theta = \int (\tan^2\theta - 2\tan\theta + 1)\,d\theta =$

   $\displaystyle\int (\sec^2\theta)\,d\theta - 2\int \tan\theta\,d\theta = \tan\theta + 2\ln(|\cos\theta| + C)$

   (b) $\displaystyle\int \sec^{\frac{3}{2}}x\tan x\,dx = \int \sec^{\frac{1}{2}}x\sec x\tan x\,dx$. Let $u = \sec x$, $du = \sec x\tan x\,dx$

   $\displaystyle\int u^{\frac{1}{2}}\,du = \dfrac{2}{3}u^{\frac{3}{2}} + C = \dfrac{2}{3}\sec^{\frac{3}{2}}x + C$

5. (a) $\dfrac{dy}{dx} = x^2\cos\dfrac{1}{x}\left(-\dfrac{1}{x^2}\right) + \sin\dfrac{1}{x}(2x) = 2x\sin\dfrac{1}{x} - \cos\dfrac{1}{x}$

   (b) Because $y = \dfrac{1}{2}\csc 2x$, then $\dfrac{dy}{dx} = \dfrac{1}{2}(-\csc 2x\cot 2x \cdot 2) = -\csc 2x\cot 2x$.

(c) $\dfrac{dy}{dx} = 2\sec\sqrt{x}\cdot\sec\sqrt{x}\tan\sqrt{x}\cdot\left(\dfrac{1}{2\sqrt{x}}\right) = \dfrac{\sec^2\sqrt{x}\tan\sqrt{x}}{\sqrt{x}}$

(d) $\dfrac{dx}{d\theta} = -3\cos^2\theta\sin\theta$ and $\dfrac{dy}{d\theta} = 3\sin^2\theta\cos\theta$. $\dfrac{dy}{dx} = \dfrac{\dfrac{dy}{d\theta}}{\dfrac{dx}{d\theta}} = \dfrac{3\sin^2\theta\cos\theta}{-3\cos^2\theta\sin\theta} = -\tan\theta$

(e) $\cos x + \sin y\dfrac{dy}{dx} = 0 \Rightarrow \dfrac{dy}{dx} = -\dfrac{\cos x}{\sin y}$

(f) $\cos(xy)\left(x\dfrac{dy}{dx} + y\right) = 1$

$x\cos(xy)\dfrac{dy}{dx} = 1 - y\cos(xy)$ and $\dfrac{dy}{dx} = \dfrac{1 - y\cos(xy)}{x\cos(xy)}$

6. $\dfrac{dy}{dx} = ac\cos ct - bc\sin ct$ and $\dfrac{d^2y}{dx^2} = -ac^2\sin ct - bc^2\cos ct = -c^2(\sin ct + b\cos ct) = -c^2y$

7. Let $u = \tan x$, then $du = \sec^2 x\,dx$ and $\displaystyle\int\dfrac{36u^2}{(6 + u^3)^2}du.$

   Let $v = 6 + u^3$, then $dv = 3u^2\,du$ and $12\displaystyle\int v^{-2}dv = \dfrac{12v^{-1}}{-1} + C = \dfrac{-12}{6 + u^3} = -\dfrac{12}{6 + \tan^3 x} + C.$

8. (a) $\displaystyle\int(\sec^4 x)\,dx = \int\sec^2 x\sec^2 x\,dx = \int(\tan^2 x + 1)\sec^2 x\,dx =$

   $\displaystyle\int\tan^2 x\sec^2 x\,dx + \int\sec^2 x\,dx \Rightarrow \dfrac{\tan^3 x}{3} + \tan x + C$

   (b) Let $u = \sqrt{x}$, $2\,du = \dfrac{dx}{\sqrt{x}}$, $\displaystyle\int\dfrac{\cos\sqrt{x}}{\sqrt{x}}dx = 2\int\cos u\,du = 2\sin\sqrt{x} + C.$

   (c) $\displaystyle\int\cos^2 x\sin^3 x\,dx = \int\cos^2 x\sin^2 x\sin x\,dx = \int\cos^2 x(1 - \cos^2 x)\sin x\,dx \Rightarrow$

   $\displaystyle\int\cos^2 x\sin x\,dx - \int\cos^4 x\sin x\,dx = -\dfrac{\cos^3 x}{3} + \dfrac{\cos^5 x}{5} + C$

Derivatives and Integrals of Exponential Functions

1. (a) $\displaystyle\int\dfrac{x + e^x}{xe^x}dx = \int\left(\dfrac{x}{xe^x} + \dfrac{e^x}{xe^x}\right)dx = \int\left(e^{-x} + \dfrac{1}{x}\right)dx = -e^{-x} + \ln x + C$

   (b) $\displaystyle\int\dfrac{\log(x^3\times 10^x)}{x}dx = \int\dfrac{\log x^3 + \log 10^x}{x}dx = \int\dfrac{3\log x + x}{x}dx = \int\left(1 + \dfrac{3}{\ln 10}\dfrac{\ln x}{x}\right)dx \Rightarrow$

   $x + \dfrac{3}{2\ln 10}(\ln x)^2 + C$

(c) Let $u = e^x - 3$, then $e^x = u + 3$ and $du = e^x dx$ and $\displaystyle\int \frac{e^{2x}}{e^x - 3}dx = \int \frac{e^x \cdot e^x dx}{e^x - 3} \Rightarrow$

$$\int \frac{(u+3)du}{u} = \int \left(1 + \frac{3}{u}\right)du = u + 3\ln|u| + C = e^x - 3 + 3\ln\left|e^x - 3\right| + C.$$

2. If $f(x) = 2^{8x^3 + 1}$, then $f'(x) = 2^{8x^3 + 1}(\ln 2)(24x^2)$ and

$f'(.5) = 2^{8(.5)^3 + 1}(\ln 2)(24(.5)^2) = 2^2(\ln 2)(6) = 24\ln 2.$

3. $f'(x) = 2be^{2bx}$ and $g'(x) = 2ae^{2ax}$. $\displaystyle\frac{dy}{dx}\left(\frac{f(x)}{g(x)}\right) = \frac{e^{2ax}2be^{2bx} - e^{2bx}2ae^{2ax}}{e^{4ax}} = \frac{e^{2ax}e^{2bx}(2b - 2a)}{e^{4ax}} =$

$\displaystyle\frac{(2b - 2a)e^{2bx}}{e^{2ax}}, \frac{f'(x)}{g'(x)} = \frac{2be^{bx}}{2ae^{ax}}.$ Therefore, $\displaystyle\frac{2be^{2bx}}{2ae^{2ax}} = \frac{(2b - 2a)e^{2bx}}{e^{2ax}}$ and $\displaystyle\frac{b}{a} = (2b - 2a)$

$\Rightarrow b = \dfrac{2a^2}{2a - 1}$

4. $f'(x) = 3ke^{kx}$ and $\dfrac{3ke^{kx}}{3e^{kx}} = -\dfrac{a}{b} \Rightarrow k = -\dfrac{a}{b}$

5. (a) Let $u = e^{2\theta}$, then $du = 2e^{2\theta}d\theta.$

$$\int e^{2\theta}\sin e^{2\theta} = \frac{1}{2}\int \sin u\, du = -\frac{1}{2}\cos u + C = -\frac{1}{2}\cos e^{2\theta} + C$$

(b) Dividing the numerator by the denominator results in

$$\int\left(e^x - \frac{e^x}{1 + e^x}\right)dx = \int e^x dx - \int \frac{e^x}{1 + e^x}dx = e^x - \ln(1 + e^x) + C.$$

(c) $e^{2\ln x} = x^2$; $\displaystyle\int e^{2\ln x}dx = \int x^2 dx = \frac{x^3}{3} + C$

Derivatives and Integrals of Logarithmic Functions

1. Because $\ln(x^2) = 2\ln x$, then $\displaystyle\int_1^{\sqrt{5}} \frac{\ln(x^2)}{x}dx = \int_1^{\sqrt{5}} \frac{2\ln x}{x}dx = 2\int_1^{\sqrt{5}} \frac{\ln x}{x}dx.$ Let $u = \ln x$ and

$du = \dfrac{dx}{x}.$ When $x = 1$, $u = \ln(1) = 0$, and when $x = \sqrt{5}$, $u = \ln(\sqrt{5})$ and

$$2 \cdot \int_0^{\ln\sqrt{5}} u\, dn = u^2\Big|_0^{\ln\sqrt{5}} = (\ln\sqrt{5})^2 - 0^2 = (\ln\sqrt{5})^2.$$

2. (a) $\dfrac{dy}{dx} = \dfrac{1}{(x + 1)(x + 2)}[(x + 1) + (x + 2)] = \dfrac{1}{x + 1} + \dfrac{1}{x + 2} = \dfrac{2x + 3}{(x + 1)(x + 2)}$

You can also use the fact that $\ln(xy) = \ln x + \ln y$ and find the derivative.

(b) If $y = -\ln\left|\dfrac{1 + \sqrt{1 - x^2}}{x}\right|$, then $y = \ln\left|\dfrac{x}{1 + \sqrt{1 - x^2}}\right| = \ln|x| - \ln\left|1 + \sqrt{1 - x^2}\right|$.

$$\frac{dy}{dx} = \frac{1}{x} - \frac{1}{1 + \sqrt{1 - x^2}} \cdot \left(\frac{1}{2}\right) \cdot (1 - x^2)^{-\frac{1}{2}} \cdot (-2x) = \frac{1}{x} + \frac{x}{(1 + \sqrt{1 - x^2})(\sqrt{1 - x^2})}$$

$$\frac{dy}{dx} = \frac{(\sqrt{1 - x^2})(1 + \sqrt{1 - x^2}) + x^2}{x(\sqrt{1 - x^2})(1 + \sqrt{1 - x^2})} = \frac{\sqrt{1 - x^2} + 1 - x^2 + x^2}{x(\sqrt{1 - x^2})(1 + \sqrt{1 - x^2})} = \frac{1}{x\sqrt{1 - x^2}}$$

3. Change the equation to the differential form, $(x + 1)dy = x(y^2 + 1)dx \Rightarrow \displaystyle\int \frac{dy}{y^2 + 1} = \int \frac{x\,dx}{x + 1}$.

Let $u = x + 1$, $du = dx$, $x = u - 1$. $\displaystyle\int \frac{dy}{y^2 + 1} = \tan^{-1} y$.

$\tan^{-1} y = x + 1 - \ln|x + 1| + C$ is the desired solution.

4. $\dfrac{dy}{dt} = \dfrac{dy}{dx} \cdot \dfrac{dx}{dt}$ and $\dfrac{dy}{dx} = 2 - \dfrac{3}{3x}$. Then $\dfrac{dy}{dt} = \left(2 - \dfrac{1}{x}\right)(-2)$. At $(1, 2)$, $\dfrac{dy}{dt} = -2$.

5. $\ln y = \ln\left(\dfrac{x + 5}{x\cos x}\right) = \ln(x + 5) - \ln x - \ln(\cos x)$

$$\frac{1}{y}\frac{dy}{dx} = \frac{1}{x + 5} - \frac{1}{x} + \frac{\sin x}{\cos x} \Rightarrow \frac{dy}{dx} = \left(\frac{-5}{x(x + 5)} + \tan x\right)y = \left[\tan x - \frac{5}{x(x + 5)}\right]\left(\frac{x + 5}{x\cos x}\right)$$

6. (a) $\displaystyle\int \frac{(x - 2)^3}{x^2}dx = \int \frac{x^3 - 6x^2 + 12x - 8}{x^2}dx =$

$$\int\left(x - 6 + \frac{12}{x} - \frac{8}{x^2}\right)dx = \frac{x^2}{2} - 6x + 12\ln|x| + \frac{8}{x} + C$$

(b) Using long division results in

$$\int\left[(x - 2) + \frac{2x + 1}{x^2 + 2x + 1}\right]dx = \frac{x^2}{2} - 2x + \int \frac{2x + 2 - 1}{(x + 1)^2}dx =$$

$$\frac{x^2}{2} - 2x + \int \frac{2x + 2}{x^2 + 2x + 1}dx - \int \frac{1}{(x + 1)^2}dx \Rightarrow \frac{x^2}{2} - 2x + \int \frac{du}{u} - \int \frac{1}{v^2}dv, \text{ where}$$

$u = x^2 + 2x + 1$, $du = (2x + 2)dx$, $v = x + 1$, $dv = dx \Rightarrow$

$$x^2 - 2x + \ln|u| + \frac{1}{v} + C = \frac{x^2}{2} - 2x + \ln\left|x^2 + 2x + 1\right| + \frac{1}{x + 1} + C.$$

$\ln\left|x^2 + 2x + 1\right| = \ln\left|(x + 1)^2\right| = 2\ln|x + 1|$ may be used as part of the answer.

(c) Let $u = 1 - \ln t$, $du + \dfrac{-1}{t}dt$. $\displaystyle\int \frac{(1 - \ln t)^2}{t}dt = -\int u^2 du = -\frac{u^3}{3} + C = -\frac{(1 - \ln t)^3}{3} + C$.

(d) Let $u = 4x - 4x^2$ and $du = (4 - 8x)dx$.

$$\Rightarrow \int \frac{2x-1}{\sqrt{4x-4x^2}}dx = -\frac{1}{4}\int \frac{du}{u^{\frac{1}{2}}} = \left(-\frac{1}{4}\right)(2)u^{\frac{1}{2}} + C = -\frac{1}{2}\sqrt{4x-4x^2} + C.$$

7. (a) $y = \ln x + \frac{1}{2}\ln(x^2 + 1)$. $\dfrac{dy}{dx} = \dfrac{1}{x} + \dfrac{1}{2} \cdot \dfrac{2x}{x^2+1} = \dfrac{2x^2+1}{x(x^2+1)}$

(b) $\dfrac{dy}{dx} = \dfrac{x(3\ln^2 x)}{x} + \ln^3 x = 3\ln^2 x + 1\ln^3 x$

(c) $\dfrac{dy}{dt} = \dfrac{1}{1-t}$ and $\dfrac{dx}{dt} = \dfrac{1}{(1-t)^2}$; $\dfrac{\frac{dy}{dt}}{\frac{dx}{dt}} = \dfrac{\frac{1}{1-t}}{\frac{1}{(1-t)^2}} = 1 - t = \dfrac{1}{x}$

8. $f(x) = 3\ln x$. $f'(x) = \dfrac{3}{x}$, $f''(x) = \dfrac{-3}{x^2}$, and $f''(3) = \dfrac{-3}{3^2} = -\dfrac{1}{3}$.

Derivatives and Integrals of Inverse Trigonometric Functions

1. $\dfrac{2}{1+4x^2} = \dfrac{2y}{y^2} \cdot \dfrac{dy}{dx}$; $\dfrac{dy}{dx} = \dfrac{y}{1+4x^2}$

2. (a) Let $\theta = \sec^{-1}x$. Using the identity $\tan^2\theta + 1 = \sec^2\theta$, then $y^2 + 1 = x^2$.

$$y = \pm\sqrt{x^2-1} = \pm(x^2-1)^{\frac{1}{2}}$$

$$\frac{dy}{dx} = \frac{\pm x}{\sqrt{x^2-1}}$$

(b) Let $u = \cos^{-1}x$; $\dfrac{dy}{dx} = \dfrac{dy}{du} \cdot \dfrac{du}{dx} = \sec^2 u\left(\dfrac{-1}{\sqrt{1-x^2}}\right) = \dfrac{1}{\cos^2 u} \cdot \left(\dfrac{-1}{\sqrt{1-x^2}}\right) = \dfrac{-1}{x^2\sqrt{1-x^2}}$.

(c) $\dfrac{dy}{dx} = \dfrac{\frac{1}{2}}{1+\frac{x^2}{4}} = \dfrac{2}{x^2+4}$

(d) $\dfrac{dy}{dx} = \dfrac{1}{\sqrt{1-x^2}} - \dfrac{1(-2x)}{2\sqrt{1-x^2}} = \dfrac{1+x}{\sqrt{1-x^2}}$

3. (a) $\displaystyle\int \frac{dy}{\sqrt{6y-y^2}} = \int \frac{dy}{\sqrt{9-(y^2-6y+9)}} = \int \frac{dy}{\sqrt{9-(y-3)^2}} = \sin^{-1}\left(\frac{y-3}{3}\right) + C$

(b) Let $u = e^x$, $du = e^x dx$, and $a = \sqrt{3}$. $\displaystyle\int \frac{e^x}{3+e^{2x}}dx = \int \frac{du}{a^2+u^2} = \frac{1}{\sqrt{3}}\tan^{-1}\frac{e^x}{\sqrt{3}} + C.$

(c) Let $u = \sqrt{x-1}$, then $u^2 = x-1$ and $2u\,du = dx$.

$$\int \frac{\sqrt{x-1}}{x}dx = 2\int \frac{u^2\,du}{u^2+1} = 2\int \frac{u^2+1-1}{u^2+1}du = 2\int \left(1 - \frac{1}{u^2+1}\right)du =$$

$$\Rightarrow 2(u - \tan^{-1}u) + C = 2(\sqrt{x-1} - \tan^{-1}\sqrt{x-1}) + C.$$

4. (a) Let $x = 2\sin\theta$ (draw a right triangle with side x opposite angle θ and hypotenuse 2).

$$dx = 2\cos\theta\,d\theta \text{ and } \sqrt{4-x^2} = 2\cos\theta. \int \frac{x^2}{\sqrt{4-x^2}}dx = \int \frac{4\sin^2\theta}{2\cos\theta}2\cos\theta\,d\theta = 4\int \sin^2\theta\,d\theta$$

$$\Rightarrow 4\int \left(\frac{1}{2} - \frac{\cos 2\theta}{2}\right)d\theta = 2\theta - \sin 2\theta + C = 2\theta - 2\sin\theta\cos\theta + C$$

$$\Rightarrow 2\sin^{-1}\frac{x}{2} - (2)\left(\frac{x}{2}\right)\left(\frac{\sqrt{4-x^2}}{2}\right) + C = 2\sin^{-1}\frac{x}{2} - \frac{x\sqrt{4-x^2}}{2} + C$$

(b) Let $2x = 3\tan\theta$, then $2dx = 3\sec^2\theta\,d\theta$ and $\sqrt{4x^2+9} = 3\sec\theta$.

$$\int \frac{dx}{x\sqrt{4x^2+9}} = \int \frac{3\sec^2\theta\,d\theta}{\frac{3\tan\theta}{2}\cdot 3\sec\theta} = \frac{1}{3}\int \frac{\sec\theta}{\tan\theta}d\theta = \frac{1}{3}\int \csc\theta\,d\theta = \frac{1}{3}\ln|\csc\theta - \cot\theta| + C$$

$$\Rightarrow \frac{1}{3}\ln\left|\frac{\sqrt{4x^2+9}-3}{2x}\right| + C.$$

The Chain Rule

1. (a) $\dfrac{dy}{dx} = -\sqrt{3}\sin(\sqrt{3}x)$

   (b) $\dfrac{dy}{dx} = (2x+7)\csc(x^2+7x)\cot(x^2+7x)$

   (c) $\dfrac{dy}{dx} = [-\sin(\sin x)]\cos x$

   (d) $\dfrac{dy}{dx} = 5\left(\frac{x}{5}+\frac{1}{5x}\right)^4\left(\frac{1}{5}+\frac{1}{5x^2}\right) = \left(\frac{x}{5}+\frac{1}{5x}\right)^4\left(1-\frac{1}{x^2}\right)$

2. (a) $\dfrac{dy}{dx} = 4x^3(2x-5)^3(2) + 3x^2(2x-5)^4 = x^2(2x-5)^3(14x-15)$

   (b) Let $y = \dfrac{\sin x}{1+\cos x}; \dfrac{dy}{dx} = \dfrac{(1+\cos x)\cos x - \sin x(-\sin x)}{(1+\cos x)^2} = \dfrac{\cos x + 1}{(1+\cos x)^2} = \dfrac{1}{1+\cos x}$

   (c) $\dfrac{dy}{dx} = 2\dfrac{1}{2\sqrt{(\csc x + \cot x)}} \cdot (-\csc x\cot x - \csc^2 x) = \dfrac{-\csc x(\cot x - \cos^2 x)}{\sqrt{\csc x + \cot x}} =$

   $-\csc x\sqrt{\csc x + \cot x}$

(d) $\frac{dy}{dx} = 3(1 + \cos^2 7x)^2 (2\cos 7x)(-7\sin 7x) = -42(\sin 7x)(\cos 7x)(1 + \cos^2 7x)^2 \Rightarrow$

$\qquad -21(\sin 14x)(1 + \cos^2 7x)^2$

3. If $2x = \sin\frac{x}{2}$, then the 2 curves intersect at the origin. For $y = \sin 2x$, $\frac{dy}{dx} = 2\cos 2x$.

   For $y = -\sin\frac{x}{2}$, $\frac{dy}{dx} = -\frac{1}{2}\cos\frac{x}{2}$. At $x = 0$, $\frac{dy}{dx} = 2$ for the first curve, and at $x = 0$, $\frac{dy}{dx} = -\frac{1}{2}$

   for the second curve. Because the derivatives are the slopes of the tangents at $x = 0$ and the slopes are negative reciprocals of each other, the tangents are perpendicular and the curves are orthogonal.

4. $\frac{dy}{dt} = 32\left(\frac{2\pi}{365}\right)\cos\left[\frac{2\pi}{365}(t - 88)\right]$. $\frac{dy}{dt}$ is largest when $t - 88 = 0$, and $t = 88$. March 29 of a

   non-leap year is the 88th day. When $t = 88$, $\frac{dy}{dt} = \frac{32(2\pi)}{365} = 0.55085\ldots$ °F/day.

Implicit Differentiation

1. For $y = 2$, $(2)^3 + x^2(2)^2 - 3x^2 = 9 \Rightarrow x^2 = 1 \Rightarrow x = \pm 1$. In the first quadrant $x = 1$

   $3y^2\frac{dy}{dx} + 2xy^2 + 2yx^2\frac{dy}{dx} - 6x = 0$

   $(3y^2 + 2yx^2)\frac{dy}{dx} = 6x - 2xy^2$

   $\frac{dy}{dx} = \frac{6x - 2xy^2}{3y^2 + 2yx^2} = \frac{6(1) - 2(1)(2)^2}{3(2)^2 + 2(2)(1)^2} = -\frac{1}{8}$

2. $6x^2 + 3x\frac{dy}{dx} + 3y + e^y\frac{dy}{dx} = 0$

   $\frac{dy}{dx} = -\frac{6x^2 + 3y}{3x + e^y}$; at $x = 0$, $y = \ln 6$, $\frac{dy}{dx} = -\frac{3\ln 6}{6} = -\frac{\ln 6}{2}$.

3. $2x - \left(x\frac{dy}{dx} + y\right) + 2y\frac{dy}{dx} = 0$

   $(2y - x)\frac{dy}{dx} = y - 2x$

   $\frac{dy}{dx} = \frac{y - 2x}{2y - x}$. at $(2a, -b)$, $\frac{dy}{dx} = \frac{-b - 2(2a)}{2(-b) - 2a} = \frac{b + 4a}{2b + 2a}$

4. $8x + 2x\frac{dy}{dx} + 2y - y^3 - 3xy^2\frac{dy}{dx} = 0$

   $(3xy^2 - 2x)\frac{dy}{dx} = 8x + 2y - y^3$

$$\frac{dy}{dx} = \frac{8x + 2y - y^3}{3xy^2 - 2x}. \text{ at } (1,-1), \; m = \frac{dy}{dx} = \frac{8 - 2 - (-1)^3}{3 \cdot 1(-1)^2 - 2 \times 1} = 7.$$

Using $(1,-1)$ and $m = 7$ for the equation of the tangent and $(1,-1)$ and $m = -\frac{1}{7}$ for the

equation of the normal, substitute into the point-slope form $y - y_1 = m(x - x_1)$.

$$y + 1 = 7(x - 1) \qquad\qquad\qquad y + 1 = \frac{-1}{7}(x - 1)$$

$7x - y - 8 = 0$ for the tangent. $x + 7y + 6 = 0$ for the normal.

Applications of the Derivative

1. Given $f(x) = \ln(x - 1)$,

    (a) $m = f'(x) = \frac{1}{x - 1}$. $f(3) = \ln 2$; $f'(3) = \frac{1}{2}$

    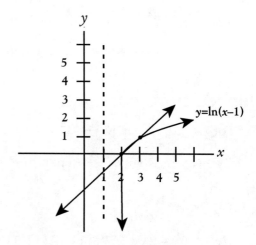

    using $y - y_1 = m(x - x_1)$; $y - \ln 2 = \frac{1}{2}(x - 3)$

    $$y = \frac{1}{2}x - \frac{3}{2} + \ln 2$$

    (b) Let $f(x) = 0$,

    $0 = \ln(x - 1)$

    $x - 1 = e^0$

    $x = 2$

    (c) Any convenient axes with the graphs of $y = \ln(x - 1)$ and $y = \frac{1}{2}x - \frac{3}{2} + \ln 2$ that show
    the points of intersection of the two graphs being approximately $(1.184, -1.714)$ and
    $(1.697, 3.041)$.

    (d) This will depend on the graphing calculator being used by the students and instructor.

2. $f'(x) = \left(\frac{2}{9}\right)\left(\frac{3}{2}\right)x^{\frac{1}{2}} = \frac{1}{3}x^{\frac{1}{2}}$

    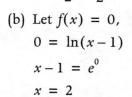

    $f'(a) = \frac{1}{3}a^{\frac{1}{2}} = \frac{5}{2}$

    $a^{\frac{1}{2}} = \frac{15}{2} \Rightarrow a = \frac{225}{4}$

3. The MVT states that the equation $f'(c) = \frac{f(b) - f(a)}{b - a}$ holds for some c between a and b.

    $a = 2$ and $b = 5$ and $f(x) = \sqrt{x - 2}$. $f'(x) = \frac{1}{2}(x - 2)^{-\frac{1}{2}} = \frac{1}{2\sqrt{x - 2}}$.

Substituting $a = 2$ and $b = 5$ and $\dfrac{f(b) - f(a)}{b - a} = \dfrac{\sqrt{3}}{3}$. $f'(c) = \dfrac{1}{2\sqrt{c - 2}}$.

Solving for c, $\dfrac{1}{2\sqrt{c - 2}} = \dfrac{\sqrt{3}}{3} \Rightarrow 12c - 24 = 9 \Rightarrow c = \dfrac{11}{4}$

4. $H'(x) = 2G'(x) + \left(-\dfrac{G'(x)}{(G(x))^2} \right)$

   $H'(0) = 2G'(0) + \dfrac{G'(0)}{(G(0))^2} = 2(-1) + \dfrac{-1}{2^2} = -\dfrac{9}{4}$

5. Because $f(-3) = f(3) = 126$ and f is differentiable, Rolle's Theorem assures the existence of at least one value of c in the interval $(-3, 3)$, where $f'(c) = 0$. $f'(x) = 8x^3 - 8x$.
   $8x^3 - 8x = 8x(x^2 - 1) = 8x(x - 1)(x + 1) = 0 \Rightarrow x = 0, x = \pm 1$ are three x-values in $(-3, 3)$ where $f'(x) = 0$.

6. At $(1, 3)$, $m = \dfrac{dy}{dx} = -2x$. The slope of the tangent line at $(1, 3)$ is -2. The equation of the tangent line is $y = -2x + 5$. The intercepts are $\dfrac{5}{2}$ and 5. The area of the triangle formed is

   $\left(\dfrac{1}{2} \right)\left(\dfrac{5}{2} \right)(5) = \dfrac{25}{4} = 6.25$ square units.

7. $m = \dfrac{dy}{dx} = 6x^2 + 3 = 9 \Rightarrow x = 1$ or $x = -1$. When $x = 1$, $y = 5$ and the tangent at $(1, 5)$ has an equation $y = 9x - 4$. At $x = -1$, $y = -5$ and the tangent at $(-1, -5)$ has an equation $y = 9x + 4$. The smallest slope is the smallest value of $\dfrac{dy}{dx}$, which is 3 when $x = 0$.

8. $\dfrac{dy}{dx} = 12x^2 - 12x - 24$. When a tangent is parallel to the x-axis, it has a slope of 0. Therefore,
   solve $\dfrac{dy}{dx} = 12x^2 - 12x - 24 = 0 = 12(x^2 - x - 2) = 12(x + 1)(x - 2) \Rightarrow x = -1$ and $x = 2$.
   The points are $(-1, 44)$ and $(2, -10)$.

9. $f'(x) = (\ln 3)3^x - 3x^2$. The function is continuous on $[0, 3]$ and is differentiable on $(0, 3)$, so the MVT applies. We need to find c such that $f'(c) = \dfrac{f(3) - f(0)}{3} = -\dfrac{1}{3}$. Using a graphing calculator, let $Y_1 = (\ln 3)3^x - 3x^2 + \dfrac{1}{3}$. Graph Y_1 and use the zero option twice to find where Y_1 has x-intercepts. $x \approx 1.2439\ldots$ and $x \approx 2.7269\ldots$.

Optimization

1. Using the distance formula to find the distance from (x, y) to $(4, 0)$,

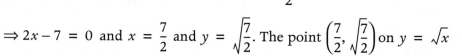

$$d = \sqrt{y^2 + (x-4)^2} = (x^2 - 7x + 16)^{\frac{1}{2}}, \quad d' = \frac{1}{2}(x^2 - 7x + 16)^{-\frac{1}{2}}(2x - 7) = 0.$$

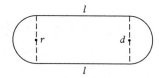

$\Rightarrow 2x - 7 = 0$ and $x = \dfrac{7}{2}$ and $y = \sqrt{\dfrac{7}{2}}$. The point $\left(\dfrac{7}{2}, \sqrt{\dfrac{7}{2}}\right)$ on $y = \sqrt{x}$

is closest to $(4, 0)$.

2. Two semicircles and two lengths of a rectangle form the perimeter of the room. The perimeter can be written $2\pi r + 2l = 200$ or

$\pi r + l = 100$. The area of the room is $A = 2rl$ and $A = 200r - 2\pi r^2$.

$A' = 200 - 4\pi r = 0$ and $r = \dfrac{50}{\pi}$ or $d = \dfrac{100}{\pi}$.

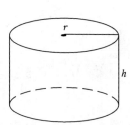

$l = 50$. The dimensions of the room are width $= \dfrac{100}{\pi}$ and length $= 50$.

3. Let a = cost for manufacturing the curved part of the cylinder and $3a$ is the cost for the bottom of the container. The total cost for the construction of the cylinder is $C = a(2\pi rh) + 3a(\pi r^2)$. The

Volume $V = \pi r^2 h = 24\pi \Rightarrow h = \dfrac{24}{r^2}$.

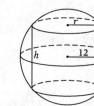

$$C = \frac{48a\pi}{r} + 3a\pi r^2 \text{ and } C' = \frac{48a\pi}{r^2} + 6a\pi r = 0$$

$\Rightarrow r^3 - 8 = 0$ and $r = 2$. The other two roots are discarded because they are complex roots. If $r = 2$, then $h = 6$, which will be the dimensions that will minimize the cost.

4. The volume of the cylinder is $V = \pi(radius)^2(height)$. The cross section of the cylinder inside the sphere provides the equation:

$$r^2 = 144 - \frac{h^2}{4} \text{ and } V = \pi\left(144h - \frac{h^3}{4}\right)$$

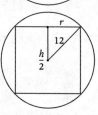

$V' = \pi\left(144 - \dfrac{3}{4}h^2\right) = 0 \Rightarrow 192 - h^2 = 0$. $h = 8\sqrt{3}$ and $h = -8\sqrt{3}$,

which is discarded.

$r = 4\sqrt{6}$. Therefore, the right circular cylinder of maximum volume that can be inscribed in a sphere of radius 12 has a height of $8\sqrt{3}$ and a radius of $4\sqrt{6}$.

5. The line passing through $A(2, 3)$ has intercepts at $X(x, 0)$ and $Y(0, y)$. The slope of $\overline{AX}$ = slope of $\overline{YX}$.

   Therefore, $\dfrac{3}{2 - x} = \dfrac{-y}{x} \Rightarrow 3x = xy - 2y \Rightarrow y = \dfrac{3x}{x - 2}$.

   Let A be the area of the right triangle formed and

   $A = \dfrac{1}{2}xy = \left(\dfrac{1}{2}\right)(x)\left(\dfrac{3x}{x - 2}\right) = \dfrac{1}{2}\left(\dfrac{3x^2}{x - 2}\right).$

   $A' = \left(\dfrac{1}{2}\right)\left(\dfrac{(x - 2)(6x) - 3x^2}{(x - 2)^2}\right) = \dfrac{3x^2 - 12x}{(x - 2)^2} = 0$

   When $x = 4$, $y = 6$. The vertices of the right triangle where the hypotenuse passes through $(2, 3)$ and intersects the coordinate axes in the first quadrant are $(0, 0)$, $(4, 0)$, and $(0, 6)$.

6. The area of the rectangle $A = xy$. Because of similar triangles,

   $\dfrac{5 - x}{y} = \dfrac{x}{12 - y}$. Solving for y, $y = 12 - \dfrac{12}{5}x$.

   $A = x\left(12 - \dfrac{12}{5}x\right) = 12x - \dfrac{12}{5}x^2$

   $A' = 12 - \dfrac{24}{5}x = 0 \Rightarrow x = \dfrac{5}{2}.$

   When $x = \dfrac{5}{2}$, $y = 6$. The dimensions of the largest rectangle that can be inscribed in a 5-12-13 right triangle with one vertex on the hypotenuse is $\dfrac{5}{2} \times 6$.

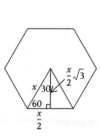

7. The area of the hexagon is $A_{hex} = 6\left(\dfrac{1}{2}\right)\left(\dfrac{x\sqrt{3}}{2}\right)(x) = \dfrac{3\sqrt{3}}{2}x^2$.

   The area of the equilateral triangle $A = \left(\dfrac{1}{2}\right)\left(\dfrac{y}{2}\sqrt{3}\right)(y) = \dfrac{\sqrt{3}}{4}y^2$

   $6x + 3y = 60 \Rightarrow 2x + y = 20 \Rightarrow y = 20 - 2x$

   Total Area $T = \dfrac{3\sqrt{3}}{2}x^2 + \dfrac{\sqrt{3}}{4}(20 - 2x)^2$

   $T' = 3\sqrt{3}x + \dfrac{\sqrt{3}}{4}(2)(20 - 2x)(-2) = 3\sqrt{3}x - 20\sqrt{3} + 2\sqrt{3}x = 0$

   $5\sqrt{3}x = 20\sqrt{3} \Rightarrow x = 4$. The length of the side of the hexagon that will provide a minimum total playing area is 4 because of $T''(4) > 0$ and the Second Derivative Test.

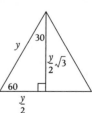

Related Rates

1. The surface area $S = 6x^2$, $\frac{dS}{dt} = 12$ in^2/sec, the volume $V = x^3$ and $\frac{dV}{dt} = 3x^2\frac{dx}{dt}$.

    Because $S = 6x^2 = 24$ in$^2 \Rightarrow x = 2$ in, $\frac{dS}{dt} = 12x\frac{dx}{dt} \Rightarrow 12$ in^2/sec $= \left(12(2)\frac{dx}{dt}\right) \Rightarrow \frac{dx}{dt} =$

    $\frac{1}{2}$ in/sec. The change in the volume of the cube, $\frac{dV}{dt} = 3(2)^2\left(\frac{1}{2}\right) = 6$ in^3/sec.

2. The volume of a hemisphere is $\frac{2}{3}\pi(radius)^3$. If the diameter is 20

    feet, the radius is 10 feet or 120 inches. Because the thickness of the
    ice is decreasing, it can be represented as

    $\frac{dx}{dt} = -0.25$ in/hr. $V = \frac{2\pi}{3}(120+x)^3 \Rightarrow \frac{dV}{dt} = 2\pi(120+x)^2\frac{dx}{dt}$

    $\frac{dV}{dt} = 2\pi(120+2)^2(-0.25) = -7,442\pi$ in^3/hr. When the ice is 2 in

    thick, its volume is decreasing at the rate of $7,442\pi$ cubic in/hr.

    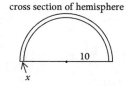

    cross section of hemisphere

    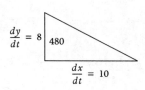

3. Given $\frac{dx}{dt} = 10$ ft/sec, $\frac{dy}{dt} = 8$ ft/sec. The girl running east runs a

    distance $x = (10 \text{ ft/sec})(120 \text{ sec}) = 1,200$ ft. The girl running north

    runs a distance of $y = (8 \text{ ft/sec})(60 \text{ sec}) = 480$ ft. The distance d they

    are apart can be represented by $d^2 = x^2 + y^2$.

    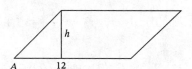

    $2d\frac{dd}{dt} = 2x\frac{dx}{dt} + 2y\frac{dy}{dt}$ and $\frac{dd}{dt} = \dfrac{x\frac{dx}{dt} + y\frac{dy}{dt}}{d}$

    $d = \sqrt{1,200^2 + 480^2} = 240\sqrt{29} \approx 1,292.44$ ft.

    $\frac{dd}{dt} = \dfrac{(1,200)(10) + (480)(8)}{240\sqrt{29}} = 12.26$ ft/sec

4. The area of the parallelogram $S = 12h$.

    $\sin A = \frac{h}{8} \Rightarrow h = 8\sin A$

    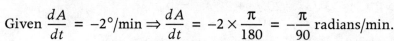

    Given $\frac{dA}{dt} = -2°$/min $\Rightarrow \frac{dA}{dt} = -2 \times \frac{\pi}{180} = -\frac{\pi}{90}$ radians/min.

    If $S = 96\sin A$, then $\frac{dS}{dt} = 96\cos A \cdot \frac{dA}{dt} = (96)(\cos 30°)\left(-\frac{\pi}{90}\right)$.

$\dfrac{dS}{dt} = (96)\left(\dfrac{\sqrt{3}}{2}\right)\left(-\dfrac{\pi}{90}\right) = \dfrac{-8\pi\sqrt{3}}{15}\text{in}^2/\text{min.}$ The area of the parallelogram is decreasing at the

rate of $\dfrac{8\pi\sqrt{3}}{15}\text{in}^2/\text{min}$ or 2.902 square in/min.

5. Ship A sailing due south is represented by $\dfrac{d\cancel{y}}{dt} = -16$ mph.

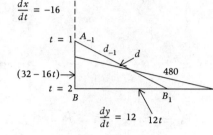

Ship B, 32 miles south of A, is represented as $\dfrac{d\cancel{x}}{dt} = 12$ mph.

The distance they are apart is represented by $d^2 = x^2 + y^2$.

$2d\dfrac{dd}{dt} = 2x\dfrac{dx}{dt} + 2y\dfrac{dy}{dt}$ and $\dfrac{dd}{dt} = \dfrac{x\dfrac{dx}{dt} + y\dfrac{dy}{dt}}{d}$. At 1 hr

or $t = 1$, $x = 16$, $y = 12$, and $d = 20$. $\dfrac{dd}{dt} = \dfrac{(16)(-16) + (12)(12)}{20} = -5.6$ mph.

Because the distance is decreasing, the ships are approaching each other. At 2 hr, $t = 2$, $x = 0$, $y = 24$, and $d = 24$.

$\dfrac{dd}{dt} = \dfrac{(0)(-16) + (24)(12)}{24} = 12$ mph. The distance is increasing, so the ships are separating.

Using $d^2 = x^2 + y^2$ and letting $x = 32 - 16t$ and $y = 12t$, then $d^2 = (32 - 16t)^2 + (12t)^2$.

$2d\dfrac{dd}{dt} = 2(32 - 16t)(-16) + 2(12t)(12) \Rightarrow d\dfrac{dd}{dt} = -512 + 256t + 144t.$

$\dfrac{dd}{dt} = \dfrac{400t - 512}{d} = 0 \Rightarrow 400t = 512 \Rightarrow t = 1.28$ hr; at $t = 1.28$ hr, $d = 19.2$ miles. Ships A

and B cease to approach each other at 1.28 hr and they are 19.2 miles apart at that time.

6. The volume of a cone is $V_c = \dfrac{1}{3}\pi r^2 h$ and the sphere is $V_s = \dfrac{4}{3}\pi r^3$.

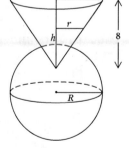

In the cone, $\dfrac{r}{h} = \dfrac{3}{4}$ and $V_c = \dfrac{3\pi}{16}h^3$; $\dfrac{dV_c}{dt} = \dfrac{9\pi}{16}h^2\dfrac{dh}{dt}$. At $h = 4$,

$\dfrac{dV_c}{dt} = 9\pi\dfrac{dh}{dt}$; $\dfrac{dV_s}{dt} = 4\pi r^2\dfrac{dr}{dt}$ and at $r = 4$, $\dfrac{dV_s}{dt} = 64\pi\dfrac{dr}{dt}$.

Because the change in volume of water running out of the cone is the same as the change in volume of water running into the sphere,

$\dfrac{dV_c}{dt} = 9\pi\dfrac{dh}{dt}$ and $9\pi\dfrac{dh}{dt} = 64\pi\dfrac{dr}{dt}$, then $\dfrac{dh}{dt} = \dfrac{64}{9}\dfrac{dr}{dt}$.

Therefore, the rate of change of the depth of the water in the cone at that instant is

approximately $\dfrac{64}{9}$ times the rate of change of the radius of the spherical balloon.

Finding Area Bounded by Functions

1. Let $3 - y^2 = y + 1 \Rightarrow y^2 + y - 2 = 0 \Rightarrow (y + 2)(y - 1) = 0$.

   $y = -2$ or $y = 1$. When $y = -2$, $x = -1$, and when $y = 1$,

   $x = 2$. Because $x = 3 - y^2$ lies to the right of $x = y + 1$, then $A =$

   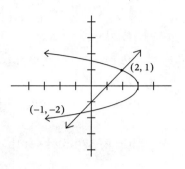

   $$\int_{-2}^{1} [(3 - y^2) - (y + 1)] \, dy = \int_{-2}^{1} (2 - y - y^2) dy$$

   $$\Rightarrow 2y - \frac{y^2}{2} - \frac{y^3}{3} \bigg|_{-2}^{1} = 4\frac{1}{2}. \text{ An alternate method would be}$$

   $$A = \int_{-1}^{2} [(x - 1) - \sqrt{3 - x}] dx + 2\int_{2}^{3} \sqrt{3 - x} \, dx.$$

2. Let $2 - x^2 = x$, then $x^2 + x - 2 = 0 \Rightarrow (x + 2)(x - 1) = 0$.

   $x = -2$ or $x = 1$. On the interval $[-2, 1]$, the graph of $f(x)$ lies above $g(x)$ and on the interval $[1, 2]$, the graph of $g(x)$ lies above $f(x)$.

   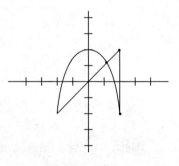

   $$A = \int_{-2}^{1} [(2 - x^2) - x] dx + \int_{1}^{2} [x - (2 - x^2)] dx$$

   $$\Rightarrow \left(2x - \frac{x^2}{2} - \frac{x^3}{3}\right)\bigg|_{-2}^{1} + \left(\frac{x^3}{3} + \frac{x^2}{2} - 2x\right)\bigg|_{1}^{2} = 4\frac{1}{2} + 1\frac{5}{6} = \frac{19}{3}$$

3. Let $A(2, -3)$, $B(6, 1)$, and $C(4, 6)$ be the vertices of $\triangle ABC$.

   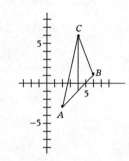

   The slope of $\overline{AC}$ is $m = \frac{9}{2}$ and its equation is $y = \frac{9}{2}x - 12$.

   The slope of $\overline{BC}$ is $m = -\frac{5}{2}$ and its equation is $y = -\frac{5}{2}x + 16$.

   The slope of $\overline{AB}$ is $m = 1$ and its equation is $y = x - 5$.

   The area A bounded by the triangle can be represented by

   $$A = \int_{2}^{4} \left[\left(\frac{9}{2}x - 12\right) - (x - 5)\right] dx + \int_{4}^{6} \left[\left(-\frac{5}{2}x + 16\right) - (x - 5)\right] dx.$$

   $$A = \left(\frac{7}{4}x^2 - 7x\right)\bigg|_{2}^{4} + \left(\frac{7}{4}x^2 + 21x\right)\bigg|_{4}^{6} = 7 + 7 = 14$$

4. $$A = \int_{1}^{9} \frac{dx}{\sqrt{x}} = \int_{1}^{9} x^{\frac{-1}{2}} dx = 2\sqrt{x}\bigg|_{1}^{9} = 4$$

   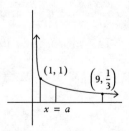

   Area of $\int_{1}^{a} \frac{dx}{\sqrt{x}} = 2\sqrt{x}\bigg|_{1}^{a} = \frac{4}{3} \Rightarrow 2\sqrt{a} - 2 = \frac{4}{3} \Rightarrow \sqrt{a} = \frac{5}{3} \Rightarrow a = \frac{25}{9}.$

5. The graphs of $y = x^{\frac{1}{m}}$ and $y = x^m$ intersect at $(0, 0)$ and $(1, 1)$.

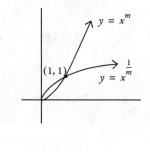

On $[0, 1]$, the graph of $y = x^{\frac{1}{m}}$ lies above $y = x^m$.
The area bounded by the two graphs is given by

$$\int_0^1 \left(x^{\frac{1}{m}} - x^m \right) dx = \frac{x^{\frac{1}{m}+1}}{\frac{1}{m}+1}\Bigg|_0^1 - \frac{x^{m+1}}{m+1}\Bigg|_0^1 = \frac{m}{1+m}\left(x^{\frac{1}{m}+1} \right)\Bigg|_0^1 - \left(\frac{1}{1+m} \right)x^{m+1}\Bigg|_0^1$$

$$\Rightarrow \left(\frac{m}{1+m} \right)(1-0) - \left(\frac{1}{1+m} \right)(1-0) = \frac{m-1}{m+1}.$$

6. Let $f(x) = g(x) \Rightarrow 10x + x^2 - 3x^3 = 2x^2 - 4x \Rightarrow 3x^3 + x^2 - 14x = 0$
   $3x^3 + x^2 - 14x = 0 \Rightarrow x(3x^2 + x - 14) = 0 \Rightarrow x(3x + 7)(x - 2) = 0.$

   Therefore, the two graphs intersect at $x = 0, 2, \dfrac{-7}{3}$.

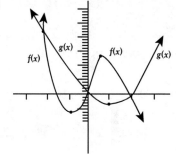

$$A = \int_{\frac{-7}{3}}^0 [g(x) - f(x)]\,dx + \int_0^2 [f(x) - g(x)]\,dx$$

$$A = \int_{\frac{-7}{3}}^0 [3x^3 + x^2 - 14x]\,dx + \int_0^2 [14x - x^2 - 3x^3]\,dx = 33.448$$

A graphing calculator expression to find the area would be as follows:

Let $y_1 = f(x)$ and $y_2 = g(x)$. Then $A = fn\,\text{int}\left(y_2 - y_1,\ x,\ \dfrac{-7}{3},\ 0 \right) + fn\text{int}(y_1 - y_2,\ x,\ 0,\ 2)$

Numerical Integration

1. The average value of f on $[a, b]$ is given by $\dfrac{1}{b-a} \displaystyle\int_a^b f(x)\,dx$.

   Using the information given, $4 = \dfrac{1}{b-2} \times 8 \Rightarrow 4(b-2) = 8 \Rightarrow b = 4.$

2. The average value of f on $[a, b]$ is given by $\dfrac{1}{b-a} \displaystyle\int_a^b f(x)\,dx$.

   Average value $= \dfrac{1}{\pi-0} \displaystyle\int_0^\pi \left(\dfrac{1}{2} + \dfrac{1}{2}\cos(2x)\right)dx = \dfrac{1}{\pi}\left(\displaystyle\int_0^\pi \dfrac{dx}{2} + \dfrac{1}{2}\displaystyle\int_0^\pi \cos(2x)dx\right)$

   Average value $= \dfrac{1}{\pi}\left(\dfrac{1}{2}x + \dfrac{1}{2}\times\dfrac{1}{2}\sin(2x)\Big|_0^\pi\right) = \dfrac{1}{\pi}\left(\dfrac{\pi}{2} + \dfrac{1}{4}(\sin 2\pi - \sin 0)\right) = \dfrac{1}{2}$

3. Average speed $=$

   $\dfrac{1}{4-2}\displaystyle\int_2^4 (5t^2 - 3t - 4)dt = \dfrac{1}{2}\left(\dfrac{5}{3}t^3 - \dfrac{3}{2}t^2 - 4t\right)\Big|_2^4 = \dfrac{1}{2}\left(\dfrac{202}{3}\right) = \dfrac{101}{3} = 33.67$ mph.

4. To use both the Trapezoidal Rule and Simpson's Rule, $h = \dfrac{b-a}{n} = \dfrac{2\pi - (-\pi)}{10} = \dfrac{3\pi}{10}$

   $T = \dfrac{h}{2}(y_0 + 2(y_1 + y_2 + y_3 + y_4 + y_5 + y_6 + y_7 + y_8 + y_9) + y_{10})$

   Let $y_0 = \dfrac{\sin(-\pi)}{-\pi}$. Increment the argument by $\dfrac{3\pi}{10}$ for each successive value of y.

   For example $y_1 = \dfrac{\sin\left(-\pi + \dfrac{3\pi}{10}\right)}{-\pi + \dfrac{3\pi}{10}}$. $T = \dfrac{3\pi}{20}(0 + 2(3.4572) + 0) = 3.258$.

   To use Simpson's rule, $h = \dfrac{3\pi}{10}$

   $S = \dfrac{h}{3}(y_0 + 4y_1 + 2y_2 + 4y_3 + 2y_4 + 4y_5 + 2y_6 + 4y_7 + 2y_8 + 4y_9 + y_{10})$

   $S = \dfrac{\pi}{10}[0 + 4(.3679) + 2(.7568) + 4(.9836) +]$ *(continued to next line)*

   $[2(.9355) + 4(.6366) + 2(.2339) + 4(-.0894) + 2(-.2162) + 4(-.1515) + 0]$

   $S = 3.270$.

5. Using the graphing calculator, the actual area is 4.821 (correct to thousandths).

   To use the Trapezoidal Rule and Simpson's Rule $h = \dfrac{b-a}{n} = \dfrac{2-0}{4} = \dfrac{1}{2}$.

   $T = \dfrac{h}{2}(y_0 + 2(y_1 + y_2 + y_3) + y_4) = \dfrac{1}{4}(2 + 2(2.031 + 2.236 + 2.716) + 3.464) = 4.858$

   $S = \dfrac{h}{3}(y_0 + 4y_1 + 2y_2 + 4y_3 + y_4) = \dfrac{1}{6}(2 + 8.124 + 4.472 + 10.864 + 3.464) = 4.821$

   Simpson's Rule compares favorably to the actual area when the calculation is correct to thousandths. The difference in the actual area and the calculation using the Trapezoidal Rule is 0.037.

6. By Simpson's Rule:

   the area $A = \dfrac{12}{3}[0 + 4(28) + 2(50) + 4(52) + 2(48) + 4(56) + 2(66) + 4(62) + 16] = 4{,}864 \text{ ft}^2$

   The amount (volume) of sand needed is $4{,}864 \text{ ft}^2 \times \dfrac{3}{4} \text{ ft} = 3{,}648 \text{ ft}^3$

   The cost is $3{,}648 \text{ ft}^3 \times \$1.65 = \$6{,}019.20$

Volumes of Solids of Revolution

1. The radius of the spherical tank is 20 ft. The cross section of the tank is a circle with equation $x^2 + y^2 = 400$. Because the oil is 15 ft deep, the volume of oil in the container is given by

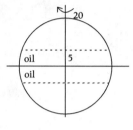

$$V = \pi \int_{5}^{20} (\sqrt{400 - y^2})^2 \, dy = \pi\left(400y - \frac{y^3}{3}\right)\Bigg|_{5}^{20} = 3{,}375\pi$$

2. Solving $x^2 = 2 - x^2 \Rightarrow x = \pm 1$. Because both graphs are symmetrical with respect to the y-axis, the solid generated by revolving the region in the first quadrant will be the same as that in the second quadrant.

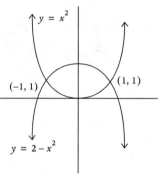

$$V = 2\pi \int_{0}^{1} (2 - x^2)^2 \, dx - 2\pi \int_{0}^{1} (x^2)^2 \, dx = 2\pi \int_{0}^{1} (4 - 4x^2) \, dx$$

$$V = 2\pi\left(4x - \frac{4}{3}x^3\right)\Bigg|_{0}^{1} = \frac{16\pi}{3}$$

3. In the first quadrant, $y = 4 - x^2$ and $y = 3x$ intersect at $(1, 3)$. The region bounded by the positive y-axis, $y = 4 - x^2$, and $y = 3x$

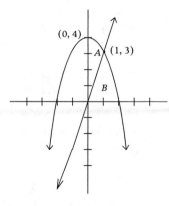

   revolved about the x-axis is $V_A = \pi \int_{0}^{1} (4 - x^2)^2 \, dx - \pi \int_{0}^{1} (3x)^2 \, dx.$

   The region bounded by the positive x-axis, $y = 3x$ and $y = 4 - x^2$

   revolved about the x-axis is $V_B = \pi \int_{0}^{1} (3x)^2 \, dx + \pi \int_{1}^{2} (4 - x^2)^2 \, dx.$

   $V_A = \dfrac{158\pi}{15}$, $V_B = \dfrac{98\pi}{15}$. Region A generates a larger volume than Region B.

4. The curves $y = x^2$ and $x = y^2$ intersect in the first quadrant at $(0, 0)$ and $(1, 1)$. In the first quadrant, the graph of $x = y^2$ lies above the graph of $y = x^2$. Another way to look at the first quadrant graphs is to say that the graph of $y = x^2$ lies to the right of $x = y^2$. Using the shell method to find the volume:

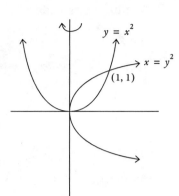

$$V = 2\pi \int_0^1 x(\sqrt{x} - x^2)\,dx = 0.94.$$ Using the disc method to find the

volume: $V = \pi \int_0^1 (\sqrt{y})^2\,dy - \pi \int_0^1 (y^2)^2\,dy = 0.94.$

5. Using the diagram at the right with the region bounded by $y = 0.25x^2$ over the interval $[0, 20]$ is revolved around the x-axis.

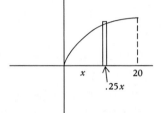

The volume can be expressed as $V = \pi \int_0^{20} (0.25x^2)^2\,dx = 40{,}000\pi$

cubic ft.

6. On the interval $[0, c]$, $2c^2 = 0 \Rightarrow c = 0$. $2c^2 = \dfrac{\pi}{2} \Rightarrow c = \dfrac{\sqrt{\pi}}{2}.$

Using the shell method, $V = 2\pi \int_0^{\frac{\sqrt{\pi}}{2}} x\cos(2x^2)\,dx$. If $u = 2x^2$, $du = 4x\,dx$.

When $x = 0$, $u = 0$. When $x = \dfrac{\sqrt{\pi}}{2}$, $u = \dfrac{\pi}{2}$. Therefore,

$V = \dfrac{\pi}{2} \int_0^{\frac{\pi}{2}} \cos u\,du = \dfrac{\pi}{2}\sin u\Big|_0^{\frac{\pi}{2}} = \dfrac{\pi}{2}.$ The definite integral expression for finding the volume by revolving the region bounded by $y = 1$ (which lies above the given function) around the x-axis is

$V = \pi \int_0^{\frac{\sqrt{\pi}}{2}} [1^2 - (\cos(2x^2))^2]\,dx = \pi \int_0^{\frac{\sqrt{\pi}}{2}} [\sin^2(2x^2)]\,dx.$ The graphing calculator expression for

finding the volume is $\pi((\sin(2x^2))^2, x, 0, \dfrac{\sqrt{\pi}}{2}) = .871.$

Areas of Surfaces of Revolution

1. Using the formula $L = \int_a^b \sqrt{\left(\dfrac{dx}{dt}\right)^2 + \left(\dfrac{dy}{dt}\right)^2}\, dt$. $\dfrac{dx}{dt} = 2t$ and $\dfrac{dy}{dt} = 3t^2$;

   substituting these values into L, $L = \int_0^2 \sqrt{(4t^2) + (9t^4)}\, dt = \int_0^2 \sqrt{4 + 9t^2}\, t\, dt$.

   Let $u = 4 + 9t^2$ and $du = 18t\, dt$ and $L = \dfrac{1}{18} \int_4^{40} u^{\frac{1}{2}}\, du = \dfrac{1}{27} u^{\frac{3}{2}} \Big|_4^{40} = 9.073$.

2. To find the surface area of a solid generated by revolving an area

   around the *y*-axis, use the formula $S = 2\pi \int_a^b x \sqrt{1 + [f'(x)]^2}\, dx$.

   Let $f(x) = 2 - 2x^2$, $f'(x) = -4x$. Substituting,

   $S = 2\pi \int_0^1 x \sqrt{1 + 16x^2}\, dx$.

   Let $u = 1 + 16x^2$, $du = 32x\, dx$.

   $S = \dfrac{\pi}{16} \int_1^{17} u^{\frac{1}{2}}\, du = \dfrac{\pi}{24} u^{\frac{3}{2}} \Big|_1^{17} = \dfrac{\pi}{24}(17\sqrt{17} - 1)$ is the exact value.

   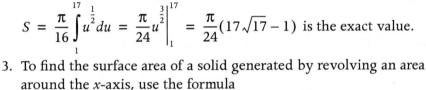

3. To find the surface area of a solid generated by revolving an area
   around the *x*-axis, use the formula

   $S = 2\pi \int_a^b f(x) \sqrt{1 + [f'(x)]^2}\, dx$. If $f(x) = \sqrt{a^2 - x^2}$,

   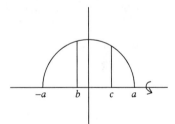

   then $f'(x) = \dfrac{-x}{\sqrt{a^2 - x^2}}$. Also $-a < b < c < a$.

   $S = 2\pi \int_b^c \sqrt{a^2 - x^2} \sqrt{1 + \left(\dfrac{-x}{\sqrt{a^2 - x^2}}\right)^2}\, dx = 2\pi \int_b^c \sqrt{a^2 - x^2} \sqrt{\dfrac{a^2}{a^2 - x^2}}\, dx$

   $S = 2\pi \int_b^c a\, dx = 2a\pi x \Big|_b^c = 2a\pi(c - b)$.

4. To find the surface area of a solid generated by revolving an area around the *y*-axis, use the formula

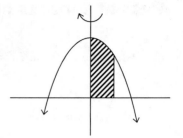

$S = 2\pi \int_a^b x\sqrt{1 + [f'(x)]^2}\,dx$. Let $f(x) = 9 - x^2$, then $f'(x) = -2x$.

Substituting, $S = 2\pi \int_0^2 x\sqrt{1 + 4x^2}\,dx$. Let $u = 1 + 4x^2$,

then $du = 8x\,dx$. When $x = 0$, $u = 1$. When $x = 2$, $u = 17$.

$$S = \frac{\pi}{4}\int_1^{17} u^{\frac{1}{2}}\,du = \frac{\pi}{6}\left(u^{\frac{3}{2}}\right)\Big|_1^{17} = \frac{\pi}{6}(17\sqrt{17} - 1).$$

Work

1. $f(x) = kx$. $F(0.2) = .2K = 100$, then $k = 500$ N/m.

$$W = \int_0^{0.8} 500x\,dx = 250x^2\Big|_0^{0.8} = 160 \text{ joule.}$$

2. $W = \int_0^6 (9 - x)(62.4)(25\pi)\,dx = 1{,}560\pi \int_0^6 (9 - x)\,dx = 1{,}560\pi\left(9x - \frac{x^2}{2}\right)\Big|_0^6 = 56{,}160\pi \text{ ft} \cdot \text{lb.}$

3. When the rocket is *x* feet above the ground, the total weight is given by total weight = rocket's weight + fuel's weight, where

$$\text{total weight} = 3 + \left[40 - 2\frac{x}{1{,}000}\right] \text{ tons}$$

$$\text{total weight} = 43 - \frac{x}{500}$$

$$Work = \int_0^{3000} \left(43 - \frac{x}{500}\right)dx = \left(43x - \frac{x^2}{1{,}000}\right)\Big|_0^{3000} = 120{,}000 \text{ ft} \cdot \text{tons}$$

Techniques of Integration

1. $\displaystyle\int \frac{dx}{2x^2 - 8x + 10} = \frac{1}{2}\int \frac{dx}{x^2 - 4x + 5} = \frac{1}{2}\int \frac{dx}{x^2 - 4x + 4 + 1} = \frac{1}{2}\int \frac{dx}{(x-2)^2 + 1} = \frac{1}{2}\tan^{-1}(x-2) + C$

2. $\displaystyle\int_0^{\frac{\pi}{4}} \sec^4 x \tan x\, dx = \int_0^{\frac{\pi}{4}} (\sec^3 x \sec x \tan x)\, dx.$ If $u = \sec x$, then $du = \sec x \tan x\, dx$. If $x = 0$, $u = 1$,

   and if $x = \dfrac{\pi}{4}$, $u = \sqrt{2}$. $\displaystyle\int_1^{\sqrt{2}} u^3\, du = \frac{u^4}{4}\Big|_1^{\sqrt{2}} = \frac{1}{4}[(\sqrt{2})^4 - 1^4] = \frac{3}{4}.$

3. Given $\displaystyle\int \frac{F\, dx}{ax^2 + bx + c} = \ln|ax^2 + bx + c| + C,$

   $\dfrac{d}{dx}(ax^2 + bx + c) = 2ax + b = F.$

4. (a) $\displaystyle\int \frac{x^2\, dx}{\sqrt{16 - x^6}}$, let $u = x^3$, $du = 3x^2\, dx$, $a = 4$, then use the form $\displaystyle\int \frac{du}{\sqrt{a^2 - u^2}} = \sin^{-1}\frac{u}{a} + C$

   $\Rightarrow \dfrac{1}{3}\sin^{-1}\dfrac{x^3}{4} + C.$

   (b) $\displaystyle\int \cot x(\ln(\sin x))\, dx$, let $u = \ln(\sin x)$, then $du = \dfrac{\cos x}{\sin x}dx = \cot x\, dx$

   $\displaystyle\int u\, du = \frac{u^2}{2} + C = \frac{(\ln(\sin x))^2}{2} + C$

   (c) $\displaystyle\int (\tan\theta + \sec\theta)^2\, d\theta = \int (\tan^2\theta + 2\tan\theta\sec\theta + \sec^2\theta)\, d\theta$

   $\Rightarrow \displaystyle\int (\sec^2\theta - 1 + 2\tan\theta\sec\theta + \sec^2\theta)\, d\theta$

   $\Rightarrow 2\displaystyle\int (\sec^2\theta)\, d\theta + 2\int \tan\theta\sec\theta\, d\theta - \int d\theta$

   $\Rightarrow 2\displaystyle\int \sec^2\theta\, d\theta + 2\int (\cos\theta)^{-2}\sin\theta\, d\theta - \int d\theta \Rightarrow 2\tan\theta + 2\sec\theta - \theta + C$

   (d) $\displaystyle\int \frac{dx}{1 + \sin x} \times \frac{1 - \sin x}{1 - \sin x} = \int \frac{1 - \sin x}{\cos^2 x}dx = \int \sec^2 x\, dx - \int \frac{\sin x}{\cos^2 x}dx = \tan x - \sec x + C$

5. $\displaystyle\int_1^2 \frac{2^{\frac{1}{x}}}{x^2}dx$ utilizes the form $\displaystyle\int a^u\, du = \frac{a^u}{\ln a} + C.$ $u = \dfrac{1}{x}$, $du = -\dfrac{dx}{x^2}$, $a = 2$, $x = 1$, $u = 1$,

   $x = 2$, $u = \dfrac{1}{2}$

   $-\displaystyle\int_1^{\frac{1}{2}} 2^u\, du = -\frac{2^u}{\ln 2}\Big|_1^{\frac{1}{2}} = \frac{-1}{\ln 2}\left(2^{\frac{1}{2}} - 2^1\right) = -\frac{\sqrt{2} - 2}{\ln 2}$

acceleration: the derivative of velocity

angle of inclination: Of a line crossing the x-axis, the smallest angle obtained measuring counterclockwise from the x-axis around the point of intersection.

antiderivative: An antiderivative of a function f is the function whose derivative is f.

common logarithms: base 10 logarithms

concave downward: refers to the graph of a differentiable function $y = f(x)$ on an interval where y' decreases.

concave upward: refers to the graph of a differentiable function $y = f(x)$ on an interval where y' increases

constant of integration: the constant C in the formula $F(x) + C$ that gives all possible antiderivatives of the function $f = \dfrac{dF}{dx}$

continuous function: A function $f(x)$ is continuous at each point c in its domain if $\lim\limits_{x \to 0} f(x) = f(c)$. A function is continuous if it is continuous at each point of its domain.

critical point: a point in the domain of a function for which the function's first derivative is zero or does not exist

derivative: the function f' derived from f for which that value at x is defined by the equation $f'(x) = \lim\limits_{\Delta x \to 0} \dfrac{f(x + \Delta x)}{\Delta x}$ whenever the limit exists

differentiable function: a function that is differentiable at every point of its domain

exponential function with base e and exponent x: $y = e^x$

implicit differentiation: the differentiation of implicitly defined functions $(f(x, y) = 0)$

indefinite integral: the set of all antiderivatives of a function

independent variable: the variable x in a function $y = f(x)$; the input variable of a function

integrand: in an integral, the function being integrated

left-hand limit: the limit of a function $f(t)$ as t approaches a number c from the left, denoted by $\lim\limits_{t \to c^-} f(t)$

limit: the number L as t approaches c of function $f(t)$

logarithm of x to the base a: the inverse of the function $y = a^x$. It exists when $a > 0$ and $a \neq 0$.

normal line: a line perpendicular to the tangent to a curve at a point of tangency

point of inflection: a point on a curve where the concavity changes

primary equation: a function that contains the quantity that is to be minimized or maximized

right-hand limit: the limit of a function $f(t)$ as t approaches a number c from the right, denoted by $\lim\limits_{t \to c^+} f(t)$

secondary equation: an equation, formula, or condition in a problem that is used to express the quantity that is being minimized or maximized as a function in one variable

slope of a curve: at a point $(a, f(a))$ on a differentiable curve $y = f(x)$, the number $f'(a)$

tangent to a curve: at a point $P(a, f(a))$, where the curve $y = f(x)$ is differentiable, the line through P with slope $f'(a)$

work: moving a mass over a distance by exerting a force

Share Your Bright Ideas with Us!

We want to hear from you! Your valuable comments and suggestions will help us meet your current and future classroom needs.

Your name_____Date_____

School name_____

School address_____

City _____State _____Zip_____Phone number (_____)_____

Grade level taught_____Subject area(s) taught_____Average class size_____

Where did you purchase this publication?_____

Was your salesperson knowledgeable about this product?   Yes_____   No_____

What monies were used to purchase this product?

____School supplemental budget   ____Federal/state funding   ____Personal

Please "grade" this Walch publication according to the following criteria:

Quality of service you received when purchasing .... A   B   C   D   F
Ease of use... A   B   C   D   F
Quality of content... A   B   C   D   F
Page layout .... A   B   C   D   F
Organization of material .... A   B   C   D   F
Suitability for grade level .... A   B   C   D   F
Instructional value.. A   B   C   D   F

COMMENTS:_____

_____

What specific supplemental materials would help you meet your current—or future—instructional needs?

_____

Have you used other Walch publications? If so, which ones?_____

May we use your comments in upcoming communications?   ____Yes   ____No

Please **FAX** this completed form to **207-772-3105**, or mail it to:

Product Development, J. Weston Walch, Publisher, P. O. Box 658, Portland, ME 04104-0658

We will send you a **FREE GIFT** as our way of thanking you for your feedback. **THANK YOU!**